Lecture Notes in Mathematics 1830

Editors:
J.–M. Morel, Cachan
F. Takens, Groningen
B. Teissier, Paris

Springer
Berlin
Heidelberg
New York
Hong Kong
London
Milan
Paris
Tokyo

Michael I. Gil'

Operator Functions and
Localization of Spectra

Springer

Author

Michael I. Gil'
Department of Mathematics
Ben Gurion University of Negev
P.O. Box 653
Beer-Sheva 84105
Israel
e-mail: gilmi@cs.bgu.ac.il

Cataloging-in-Publication Data applied for
Bibliographic information published by Die Deutsche Bibliothek

Die Deutsche Bibliothek lists this publication in the Deutsche Nationalbibliografie;
detailed bibliographic data is available in the Internet at http://dnb.ddb.de

Mathematics Subject Classification (2000): 47A10, 47A55, 47A56, 47A75, 47E05, 47G10, 47G20, 30C15, 45P05, 15A09, 15A18, 15A42

ISSN 0075-8434
ISBN 3-540-2246-3 Springer-Verlag Berlin Heidelberg New York

Springer-Verlag is a part of Springer Science+Business Media

springeronline.com

© Springer-Verlag Berlin Heidelberg 2003
Printed in Germany

Typesetting: Camera-ready TEX output by the authors

SPIN: 10964781   41/3142/du - 543210 - Printed on acid-free paper

Preface

1. A lot of books and papers are concerned with the spectrum of linear operators but deal mainly with the asymptotic distributions of the eigenvalues. However, in many applications, for example, in numerical mathematics and stability analysis, bounds for eigenvalues are very important, but they are investigated considerably less than asymptotic distributions. The present book is devoted to the spectrum localization of linear operators in a Hilbert space. Our main tool is the estimates for norms of operator-valued functions. One of the first estimates for the norm of a regular matrix-valued function was established by I. M. Gel'fand and G. E. Shilov in connection with their investigations of partial differential equations, but this estimate is not sharp; it is not attained for any matrix. The problem of obtaining a precise estimate for the norm of a matrix-valued function has been repeatedly discussed in the literature. In the late 1970s, I obtained a precise estimate for a regular matrix-valued function. It is attained in the case of normal matrices. Later, this estimate was extended to various classes of nonselfadjoint operators, such as Hilbert-Schmidt operators, quasi-Hermitian operators (i.e., linear operators with completely continuous imaginary components), quasiunitary operators (i.e., operators represented as a sum of a unitary operator and a compact one), etc. Note that singular integral operators and integro-differential ones are examples of quasi-Hermitian operators.

On the other hand, Carleman, in the 1930s, obtained an estimate for the norm of the resolvent of finite dimensional operators and of operators belonging to the Neumann-Schatten ideal. In the early 1980s sharp estimates for norms of the resolvent of nonselfadjoint operators of various types were established, that supplement and extend Carleman's estimates. In this book, we present the mentioned estimates and, as it was pointed out, systematically apply them to spectral problems.

2. The book consists of 19 chapters. In Chapter 1, we present some well-known results for use in the next chapters.

Chapters 2-5 of the book are devoted to finite dimensional operators and functions of such operators.

In Chapter 2 we derive estimates for the norms of operator-valued functions in a Euclidean space. In addition, we prove relations for eigenvalues of finite matrices, which improve Schur's and Brown's inequalities.

Although excellent computer softwares are now available for eigenvalue computation, new results on invertibility and spectrum inclusion regions for finite matrices are still important, since computers are not very useful, in particular, for analysis of matrices dependent on parameters. But such matrices play an essential role in various applications, for example, in the stability and boundedness of coupled systems of partial differential equations. In addition, the bounds for eigenvalues of finite matrices allow us to derive the bounds for spectra of infinite matrices. Because of this, the problem of finding invertibility conditions and spectrum inclusion regions for finite matrices continues to attract the attention of many specialists. Chapter 3 deals with various invertibility conditions. In particular, we improve the classical Levy-Desplanques theorem and other well-known invertibility results for matrices that are close to triangular ones. Chapter 4 is concerned with perturbations of finite matrices and bounds for their eigenvalues. In particular, we derive upper and lower estimates for the spectral radius. Under some restrictions, these estimates improve the Frobenius inequalities. Moreover, we present new conditions for the stability of matrices, which supplement the Rohrbach theorem.

Chapter 5 is devoted to block matrices. In this chapter, we derive the invertibility conditions, which supplement the generalized Hadamard criterion and some other well-known results for block matrices.

Chapters 6-9 form the crux of the book. Chapter 6 contains the estimates for the norms of the resolvents and analytic functions of compact operators in a Hilbert space. In particular, we consider Hilbert-Schmidt operators and operators belonging to the von Neumann-Schatten ideals.

Chapter 7 is concerned with the estimates for the norms of resolvents and analytic functions of non-compact operators in a Hilbert space. In particular, we consider so-called P-triangular operators. Roughly speaking, a P-triangular operator is a sum of a normal operator and a compact quasinilpotent one, having a sufficiently rich set of invariant subspaces. Operators having compact Hermitian components are examples of P-triangular operators.

In Chapters 8 and 9 we derive the bounds for the spectra of quasi-Hermitian operators.

In Chapter 10 we introduce the notion of the multiplicative operator integral. By virtue of the multiplicative operator integral, we derive spectral representations for the resolvents of various linear operators. That representation is a generalization of the classical spectral representation for resolvents of normal operators. In the corresponding cases the multiplicative integral is an operator product.

Chapters 11 and 12 are devoted to perturbations of the operators of the form $A = D + W$, where D is a normal boundedly invertible operator and $D^{-1}W$ is compact. In particular, estimates for the resolvents and bounds for the spectra are established.

Chapters 13 and 14 are concerned with applications of the main results from Chapters 7-12 to integral, integro-differential and differential operators, as well as to infinite matrices. In particular, we suggest new estimates for the spectral radius of integral operators and infinite matrices. Under some restrictions, they improve the classical results.

Chapter 15 deals with operator matrices. The spectrum of operator matrices and related problems have been investigated in many works. Mainly, Gershgorin-type bounds for spectra of operator matrices with bounded operator entries are derived. But Gershgorin-type bounds give good results in the cases when the diagonal operators are dominant. In Chapter 15, under some restrictions, we improve these bounds for operator matrices. Moreover, we consider matrices with unbounded operator entries. The results of Chapter 15 allow us to derive bounds for the spectra of matrix differential operators.

Chapters 16-18 are devoted to Hille-Tamarkin integral operators and matrices, as well as integral operators with bounded kernels.

Chapter 19 is devoted to applications of our abstract results to the theory of finite order entire functions. In that chapter we consider the following problem: if the Taylor coefficients of two entire functions are close, how close are their zeros? In addition, we establish bounds for sums of the absolute values of the zeros in the terms of the coefficients of its Taylor series. They supplement the Hadamard theorem.

3. This is the first book that presents a systematic exposition of bounds for the spectra of various classes of linear operators in a Hilbert space. It is directed not only to specialists in functional analysis and linear algebra, but to anyone interested in various applications who has had at least a first year graduate level course in analysis. The functional analysis is developed as needed.

I was very fortunate to have had fruitful discussions with the late Professors I.S. Iohvidov and M.A. Krasnosel'skii, to whom I am very grateful for their interest in my investigations.

Michael I. Gil'

Table of Contents

1. Preliminaries

In this chapter we present some well-known results for use in the next chapters.

1.1   Vector and Matrix Norms

Let $\mathbf{C}^n$ be an n-dimensional complex Euclidean space. A function

$$\nu : \mathbf{C}^n \to [0, \infty)$$

is said to be *a norm on* $\mathbf{C}^n$ (or a vector norm), if ν satisfies the following conditions:

$$\nu(x) = 0 \text{ iff } x = 0, \nu(\alpha x) = |\alpha|\nu(x), \ \nu(x+y) \le \nu(x) + \nu(y) \qquad (1.1)$$

for all $x, y \in \mathbf{C}^n$, $\alpha \in \mathbf{C}$. Usually, a norm is denoted by the symbol $\|.\|$. That is, $\nu(x) = \|x\|$. The following important properties follow immediately from the definition:

$$\|x - y\| \ge \|x\| - \|y\| \text{ and } \|x\| = \| - x\|.$$

There are an infinite number of norms on $\mathbf{C}^n$. However, the following norms are most commonly used in practice:

$$\|x\|_p = [\sum_{k=1}^{n} |x_k|^p]^{1/p} \ (1 \le p < \infty) \text{ and } \|x\|_\infty = \max_{k=1,\dots,n} |x_k|$$

for an $x = (x_k) \in \mathbf{C}^n$. The norm $\|x\|_2$ is called *the Euclidean norm*.

   Throughout this chapter $\|x\|$ *means an arbitrary norm of a vector* x. We will use the following matrix norms: the operator norm and the Frobenius

(Hilbert-Schmidt) norm. The operator norm of a matrix (a linear operator in $\mathbf{C}^n$) A is

$$\|A\| = \sup_{x \in C^n} \frac{\|Ax\|}{\|x\|}.$$

The relations

$$\|A\| > 0 \ (A \neq 0), \ \|\lambda A\| = |\lambda| \|A\| \ (\lambda \in \mathbf{C}),$$

$$\|AB\| \leq \|A\| \|B\|, \text{ and } \|A + B\| \leq \|A\| + \|B\|$$

are valid for all matrices A and B. The Frobenius norm of A is

$$N(A) = \sqrt{\sum_{j,k=1}^{n} |a_{jk}|^2}.$$

Here a_{jk} are the entries of matrix A in some orthogonal normal basis. The Frobenius norm does not depend on the choice of an orthogonal normal basis. The relations

$$N(A) > 0 \ (A \neq 0); \ N(\lambda A) = |\lambda| N(A) \ (\lambda \in \mathbf{C}),$$

$$N(AB) \leq N(A)N(B) \text{ and } N(A + B) \leq N(A) + N(B)$$

are true for all matrices A and B.

1.2 Classes of Matrices

For an $n \times n$-matrix A, A^* denotes the conjugate matrix. That is, if a_{jk} are entries of A, then $\bar{a}_{kj}$ $(j, k = 1, ..., n)$ are entries of A^*. In other words

$$(Ax, y) = (x, A^*y) \ (x, y \in \mathbf{C}).$$

The symbol $(.,.) = (.,.)_{C^n}$ means the scalar product in $\mathbf{C}^n$. We use I to denote the unit matrix in $\mathbf{C}^n$.

Definition 1.2.1 *A matrix $A = (a_{jk})_{j,k=1}^{n}$ is*
1. *symmetric (Hermitian) if $A^* = A$;*
2. *positive definite (negative definite) if it is Hermitian and*

$$(Ah, h) \geq (\leq) 0 \ (h \in \mathbf{C}^n);$$

3. *unitary if $A^*A = AA^* = I$;*
4. *normal if $AA^* = A^*A$;*
5. *nilpotent if $A^n = 0$.*

Let A be an arbitrary matrix. Then the matrices

$$A_I = (A - A^*)/2i \text{ and } A_R = (A + A^*)/2$$

are the imaginary Hermitian component and the real Hermitian one of A, respectively. A matrix A *is dissipative* if its real Hermitian component is negative definite. By A^{-1} the matrix inverse to A is denoted: $AA^{-1} = A^{-1}A = I$.

1.3   Eigenvalues of Matrices

Let A be an arbitrary matrix. Then if for some $\lambda \in \mathbf{C}$, the equation $Ah = \lambda h$ has a nontrivial solution, λ is an eigenvalue of A and h is its eigenvector. An eigenvalue λ has the (algebraic) multiplicity r if

$$dim(\cup_{k=1}^{n} ker(A - \lambda I)^k) = r.$$

Here $ker\ B$ denotes the kernel of a mapping B.

Let $\lambda_k(A)$ $(k = 1, \ldots, n)$ be the eigenvalues of A, including with their multiplicities. Then the set $\sigma(A) = \{\lambda_k(A)\}_{k=1}^{n}$ is *the spectrum* of A.

All the eigenvalues of a Hermitian matrix A are real. If, in addition, A is positive (negative) definite, then all its eigenvalues are non-negative (non-positive). Furthermore,

$$r_s(A) = \max_{k=1,\ldots,n} |\lambda_k(A)|$$

is the spectral radius of A. Denote

$$\alpha(A) = \max_{k=1,\ldots,n} Re\lambda_k(A), \ \beta(A) = \min_{k=1,\ldots,n} Re\lambda_k(A).$$

A matrix A is said to be *a Hurwitz matrix* if all its eigenvalues lie in the open left half-plane, i.e., $\alpha(A) < 0$.

A complex number λ is *a regular point of A* if it does not belong to the spectrum of A, i.e., if $\lambda \neq \lambda_k(A)$ for any $k = 1, \ldots, n$.

The trace of A is sometimes denoted by $Tr\ (A)$:

$$Trace\ (A) = Tr\ (A) = \sum_{k=1}^{n} \lambda_k(A).$$

So the Frobenius norm can be defined as

$$N^2(A) = Trace\ (A^*A) = Trace\ (AA^*).$$

Recall that $Tr(AB) = Tr(BA)$ and $Tr(A + B) = Tr(A) + Tr(B)$ for all matrices A and B. In addition, $det(A)$ means the determinant of A:

$$det(A) = \prod_{k=1}^{n} \lambda_k(A).$$

The polynomial

$$p(\lambda) = det(\lambda I - A) = \prod_{k=1}^{n} (\lambda - \lambda_k(A))$$

is said to be the characteristic polynomial of A. All the eigenvalues of A are the roots of its characteristic polynomial. The algebraic multiplicity of an eigenvalue of A coincides with the multiplicity of the corresponding root of the characteristic polynomial. A polynomial is said to be *a Hurwitz one* if all its roots lie in the open left half-plane. Thus, the characteristic polynomial of a Hurwitz matrix is a Hurwitz polynomial.

1.4   Matrix-Valued Functions

Let A be a matrix and let $f(\lambda)$ be a scalar-valued function which is analytical on a neighborhood M of $\sigma(A)$. We define the function $f(A)$ of A by the generalized integral formula of Cauchy

$$f(A) = -\frac{1}{2\pi i} \int_\Gamma f(\lambda) R_\lambda(A) d\lambda, \qquad (4.1)$$

where $\Gamma \subset M$ is a closed smooth contour surrounding $\sigma(A)$, and

$$R_\lambda(A) = (A - \lambda I)^{-1}$$

is the resolvent of A. If an analytic function $f(\lambda)$ is represented in some domain by the Taylor series

$$f(\lambda) = \sum_{k=0}^\infty c_k \lambda^k,$$

and the series

$$\sum_{k=0}^\infty c_k A^k$$

converges in the norm of space $\mathbf{C}^n$, then

$$f(A) = \sum_{k=0}^\infty c_k A^k.$$

In particular, for any matrix A,

$$e^A = \sum_{k=0}^\infty \frac{A^k}{k!}.$$

Example 1.4.1 *Let A be a diagonal matrix:*

$$A = \begin{pmatrix} a_1 & 0 & \dots & 0 \\ 0 & a_2 & \dots & 0 \\ . & \dots & . & . \\ 0 & \dots & 0 & a_n \end{pmatrix}.$$

Then

$$f(A) = \begin{pmatrix} f(a_1) & 0 & \dots & 0 \\ 0 & f(a_2) & \dots & 0 \\ . & \dots & . & . \\ 0 & \dots & 0 & f(a_n) \end{pmatrix}.$$

Example 1.4.2 *If a matrix J is an $n \times n$-Jordan block:*

$$J = \begin{pmatrix} \lambda_0 & 1 & 0 & \ldots & 0 \\ 0 & \lambda_0 & 1 & \ldots & 0 \\ \cdot & \cdot & \cdot & \ldots & \cdot \\ \cdot & \cdot & \cdot & \ldots & \cdot \\ \cdot & \cdot & \cdot & \ldots & \cdot \\ 0 & 0 & \ldots & \lambda_0 & 1 \\ 0 & 0 & \ldots & 0 & \lambda_0 \end{pmatrix},$$

then

$$f(J) = \begin{pmatrix} f(\lambda_0) & \frac{f'(\lambda_0)}{1!} & \ldots & \frac{f^{(n-1)}(\lambda_0)}{(n-1)!} \\ 0 & f(\lambda_0) & \ldots & \\ \cdot & \cdot & \ldots & \cdot \\ \cdot & \cdot & \ldots & \cdot \\ \cdot & \cdot & \ldots & \cdot \\ 0 & \ldots & f(\lambda_0) & \frac{f'(\lambda_0)}{1!} \\ 0 & \ldots & 0 & f(\lambda_0) \end{pmatrix}.$$

1.5   Contour Integrals

Lemma 1.5.1 *Let M_0 be the closed convex hull of points $x_0, x_1, ..., x_n \in \mathbf{C}$ and let a scalar-valued function f be regular on a neighborhood D_1 of M_0. In addition, let $\Gamma \subset D_1$ be a Jordan closed contour surrounding the points $x_0, x_1, ..., x_n$. Then*

$$\left| \frac{1}{2\pi i} \int_\Gamma \frac{f(\lambda)d\lambda}{(\lambda - x_0)...(\lambda - x_n)} \right| \le \frac{1}{n!} \sup_{\lambda \in M_0} |f^{(n)}(\lambda)|.$$

Proof: First, let all the points be distinct: $x_j \ne x_k$ for $j \ne k$ ($j, k = 0, ..., n$), and let $D_f(x_0, x_1, ..., x_n)$ be a divided difference of function f at points $x_0, x_1, ..., x_n$. The divided difference admits the representation

$$D_f(x_0, x_1, ..., x_n) = \frac{1}{2\pi i} \int_\Gamma \frac{f(\lambda)d\lambda}{(\lambda - x_0)...(\lambda - x_n)} \qquad (5.1)$$

(see (Gel'fond, 1967, formula (54))). But, on the other hand, the following estimate is well-known:

$$| D_f(x_0, x_1, ..., x_n) | \le \frac{1}{n!} \sup_{\lambda \in M_0} |f^{(n)}(\lambda)|$$

(Gel'fond, 1967, formula (49)). Combining that inequality with relation (5.1), we arrive at the required result. If $x_j = x_k$ for some $j \ne k$, then the claimed inequality can be obtained by small perturbations and the previous reasonings. $\square$

Lemma 1.5.2 *Let $x_0 \leq x_1 \leq ... \leq x_n$ be real points and let a function f be regular on a neighborhood D_1 of the segment $[x_0, x_n]$. In addition, let $\Gamma \subset D_1$ be a Jordan closed contour surrounding $[x_0, x_n]$. Then there is a point $\eta \in [x_0, x_n]$, such that the equality*

$$\frac{1}{2\pi i} \int_\Gamma \frac{f(\lambda)d\lambda}{(\lambda - x_0)...(\lambda - x_n)} = \frac{1}{n!} f^{(n)}(\eta)$$

is true.

Proof: First suppose that all the points are distinct: $x_0 < x_1 < ... < x_n$. Then the divided difference $D_f(x_0, x_1, ..., x_n)$ of f in the points $x_0, x_1, ..., x_n$ admits the representation

$$D_f(x_0, x_1, ..., x_n) = \frac{1}{n!} f^{(n)}(\eta)$$

with some point $\eta \in [x_0, x_n]$ (Gel'fond, 1967, formula (43)), (Ostrowski, 1973, page 5). Combining that equality with representation (5.1), we arrive at the required result. If $x_j = x_k$ for some $j \neq k$, then the claimed inequality can be obtained by small perturbations and the previous reasonings. $\square$

1.6 Algebraic Equations

Let us consider the algebraic equation

$$z^n = P(z) \quad (n > 1), \text{ where } P(z) = \sum_{j=0}^{n-1} c_j z^{n-j-1} \tag{6.1}$$

with non-negative coefficients c_j $(j = 0, ..., n - 1)$.

Lemma 1.6.1 *The extreme right-hand root z_0 of equation (6.1) is non-negative and the following estimates are valid:*

$$z_0 \leq [P(1)]^{1/n} \text{ if } P(1) \leq 1, \tag{6.2}$$

and

$$1 \leq z_0 \leq P(1) \text{ if } P(1) \geq 1. \tag{6.3}$$

Proof: Since all the coefficients of $P(z)$ are non-negative, it does not decrease as $z > 0$ increases. From this it follows that if $P(1) \leq 1$, then $z_0 \leq 1$. So $z_0^n \leq P(1)$, as claimed.

Now let $P(1) \geq 1$, then due to (6.1) $z_0 \geq 1$, because $P(z)$ does not decrease. It is clear that

$$P(z_0) \leq z_0^{n-1} P(1)$$

in this case. Substituting this inequality in (6.1), we get (6.3). $\square$

Setting $z = ax$ with a positive constant a in (6.1), we obtain

$$x^n = \sum_{j=0}^{n-1} c_j a^{-j-1} x^{n-j-1}. \tag{6.4}$$

Let

$$a \equiv 2 \max_{j=0,\ldots,n-1} {}^{j+1}\!\sqrt{c_j}.$$

Then

$$\sum_{j=0}^{n-1} c_j a^{-j-1} \leq \sum_{j=0}^{n-1} 2^{-j-1} = 1 - 2^{-n+1} < 1.$$

Let x_0 be the extreme right-hand root of equation (6.4), then by (6.2) we have $x_0 \leq 1$. Since $z_0 = ax_0$, we have derived

Corollary 1.6.2 *The extreme right-hand root z_0 of equation (6.1) is non-negative. Moreover,*

$$z_0 \leq 2 \max_{j=0,\ldots,n-1} {}^{j+1}\!\sqrt{c_j}.$$

1.7 The Triangular Representation of Matrices

Let $B(\mathbf{C}^n)$ be the set of all linear operators (matrices) in $\mathbf{C}^n$. A subspace $M \subset \mathbf{C}^n$ is *an invariant subspace* of an $A \in B(\mathbf{C}^n)$, if the relation $h \in M$ implies $Ah \in M$. If P is a projector onto an invariant subspace of A, then

$$PAP = AP. \tag{7.1}$$

By Schur's theorem (Marcus and Minc, 1964, Section I.4.10.2), for a linear operator $A \in B(\mathbf{C}^n)$, there is an orthogonal normal basis $\{e_k\}$, such that A is a triangular matrix. That is,

$$Ae_k = \sum_{j=1}^{k} a_{jk} e_j \text{ with } a_{jk} = (Ae_k, e_j) \ (j = 1, \ldots, n), \tag{7.2}$$

where $(.,.)$ is the scalar product. This basis is called *Schur's basis* of the operator A. In addition,

$$a_{jj} = \lambda_j(A),$$

where $\lambda_j(A)$ are the eigenvalues of A. According to (7.2),

$$A = D + V \tag{7.3}$$

with a normal (diagonal) operator D defined by

$$De_j = \lambda_j(A)e_j \ (j = 1, \ldots, n)$$

and a nilpotent (upper-triangular) operator V defined by

$$Ve_k = \sum_{j=1}^{k-1} a_{jk}e_j \quad (k = 2, ..., n).$$

We will call equality (7.3) *the triangular representation* of matrix A. In addition, D and V will be called *the diagonal part* and *the nilpotent part* of A, respectively.

Put

$$P_j = \sum_{k=1}^{j} (., e_k)e_k \quad (j = 1, ..., n), \quad P_0 = 0.$$

Then

$$0 \subset P_1 \mathbf{C}^n \subset ... \subset P_n \mathbf{C}^n = \mathbf{C}^n.$$

Moreover,

$$AP_k = P_k AP_k; \quad VP_{k-1} = P_k VP_k; \quad DP_k = DP_k \ (k = 1, ..., n). \tag{7.4}$$

So A, V and D have the same chain of invariant subspaces.

Lemma 1.7.1 *Let $Q, V \in B(\mathbf{C}^n)$ and let V be a nilpotent operator. Suppose that all the invariant subspaces of V and of Q are the same. Then VQ and QV are nilpotent operators.*

Proof: Since all the invariant subspaces of V and Q are the same, these operators have the same basis of the triangular representation. Taking into account that the diagonal entries of V are equal to zero, we easily determine that the diagonal entries of QV and VQ are equal to zero. This proves the required result. □

1.8   Notes

This book presupposes a knowledge of basic matrix theory, for which there are good introductory texts. The books (Bellman, 1970), (Gantmaher, 1967), (Marcus and Minc, 1964) are classical. For more details about the notions presented in Sections 1.1-1.4 also see (Collatz, 1966) and (Stewart and Sun, 1990).

Estimates for roots of algebraic equations similar to Corollary 1.6.2 can be found in (Ostrowski, 1973, page 277).

References

[1] Bellman, R.E. (1970). *Introduction to Matrix Analysis*. McGraw-Hill, New York.

[2] Collatz, L. (1966). *Functional Analysis and Numerical Mathematics.* Academic Press, New York-London.

[3] Gantmaher, F. R. (1967). *Theory of Matrices.* Nauka, Moscow (In Russian).

[4] Gelfond, A. O. (1967). *Calculations of Finite Differences.* Nauka, Moscow (In Russian).

[5] Marcus, M. and Minc, H. (1964). *A Survey of Matrix Theory and Matrix Inequalities.* Allyn and Bacon, Boston .

[6] Ostrowski, A. M. (1973). *Solution of Equations in Euclidean and Banach spaces.* Academic Press, New York - London.

[7] Stewart, G. W. and Sun Ji-guang (1990). *Matrix Perturbation Theory.* Academic Press, New York.

2. Norms of Matrix-Valued Functions

In the present chapter we derive estimates for the norms of operator-valued functions in a Euclidean space. In addition, we prove relations for eigenvalues of finite matrices, which improve Schur's and Brown's inequalities.

2.1 Estimates for the Euclidean Norm of the Resolvent

Throughout the present chapter $\|.\|$ means the Euclidean norm. That is, $\|.\| = \|.\|_2$ (see Section 1.1).

Let $A = (a_{jk})$ be an $n \times n$-matrix $(n > 1)$. *The following quantity plays a key role in the sequel:*

$$g(A) = (N^2(A) - \sum_{k=1}^{n} |\lambda_k(A)|^2)^{1/2}. \tag{1.1}$$

Recall that I is the unit matrix, $N(A)$ is the Frobenius (Hilbert-Schmidt) norm of A, and $\lambda_k(A)$ $(k = 1, ..., n)$ are the eigenvalues taken with their multiplicities. Since

$$\sum_{k=1}^{n} |\lambda_k(A)|^2 \geq |Trace\ A^2|,$$

we get

$$g^2(A) \leq N^2(A) - |Trace\ A^2|. \tag{1.2}$$

In Section 2.2 we will prove the relations

$$g^2(A) \leq \frac{1}{2} N^2(A^* - A) \tag{1.3}$$

and

$$g(e^{i\tau} A + zI) = g(A) \qquad (1.4)$$

for all $\tau \in \mathbf{R}$ and $z \in \mathbf{C}$. To formulate the result, for a natural $n > 1$ introduce the numbers

$$\gamma_{n,k} = \sqrt{\frac{C_{n-1}^k}{(n-1)^k}} \quad (k = 1, ..., n-1) \text{ and } \gamma_{n,0} = 1.$$

Here

$$C_{n-1}^k = \frac{(n-1)!}{(n-k-1)!k!}$$

are binomial coefficients. Evidently, for all $n > 2$,

$$\gamma_{n,k}^2 = \frac{(n-1)(n-2)\dots(n-k)}{(n-1)^k k!} \le \frac{1}{k!} \quad (k = 1, 2, ..., n-1). \qquad (1.5)$$

Theorem 2.1.1 *Let A be a linear operator in $\mathbf{C}^n$. Then its resolvent $R_\lambda(A) = (A - \lambda I)^{-1}$ satisfies the inequality*

$$\|R_\lambda(A)\| \le \sum_{k=0}^{n-1} \frac{g^k(A)\gamma_{n,k}}{\rho^{k+1}(A,\lambda)} \text{ for any regular point } \lambda \text{ of } A,$$

where $\rho(A, \lambda) = \min_{k=1,...,n} |\lambda - \lambda_k(A)|$.

The proof of this theorem is divided into a series of lemmas which are presented in Sections 2.3-2.6.

Theorem 2.1.1 is exact: if A is a normal matrix, then $g(A) = 0$ and

$$\|R_\lambda(A)\| = \frac{1}{\rho(A, \lambda)} \text{ for all regular points } \lambda \text{ of } A.$$

Let A be an invertible $n \times n$-matrix. Then by Theorem 2.1.1,

$$\|A^{-1}\| \le \sum_{k=0}^{n-1} g^k(A) \frac{\gamma_{n,k}}{\rho_0^{k+1}(A)},$$

where $\rho_0(A) = \rho(A, 0)$ is the smallest modulus of the eigenvalues of A:

$$\rho_0(A) = \inf_{k=1,...,n} |\lambda_k(A)|.$$

Moreover, Theorem 2.1.1 and inequalities (1.5) imply

Corollary 2.1.2 *Let A be a linear operator in $\mathbf{C}^n$. Then*

$$\|R_\lambda(A)\| \le \sum_{k=0}^{n-1} \frac{g^k(A)}{\sqrt{k!}\rho^{k+1}(A,\lambda)} \text{ for any regular point } \lambda \text{ of } A.$$

An independent proof of this corollary is presented in Section 2.6.

2.2 Examples

In this section we present some examples of calculations of $g(A)$.

Example 2.2.1 *Consider the matrix*

$$A = \begin{pmatrix} a_{11} & a_{12} \\ a_{21} & a_{22} \end{pmatrix}$$

where a_{jk} $(j, k = 1, 2)$ are real numbers.

First, consider the case of *nonreal eigenvalues*: $\lambda_2(A) = \overline{\lambda}_1(A)$. It can be written

$$det(A) = \lambda_1(A)\overline{\lambda}_1(A) = |\lambda_1(A)|^2$$

and

$$|\lambda_1(A)|^2 + |\lambda_2(A)|^2 = 2|\lambda_1(A)|^2 = 2det(A) = 2[a_{11}a_{22} - a_{21}a_{12}].$$

Thus,

$$g^2(A) = N^2(A) - |\lambda_1(A)|^2 - |\lambda_2(A)|^2 =$$
$$a_{11}^2 + a_{12}^2 + a_{21}^2 + a_{22}^2 - 2[a_{11}a_{22} - a_{21}a_{12}].$$

Hence,

$$g(A) = \sqrt{(a_{11} - a_{22})^2 + (a_{21} + a_{12})^2}. \tag{2.1}$$

Let $n = 2$ and a matrix A have real entries again, but now the *eigenvalues of A are real*. Then

$$|\lambda_1(A)|^2 + |\lambda_2(A)|^2 = Trace\ A^2.$$

Obviously,

$$A^2 = \begin{pmatrix} a_{11}^2 + a_{12}a_{21} & a_{11}a_{12} + a_{12}a_{22} \\ a_{21}a_{11} + a_{21}a_{22} & a_{22}^2 + a_{21}a_{12} \end{pmatrix}.$$

We thus get the relation

$$|\lambda_1(A)|^2 + |\lambda_2(A)|^2 = a_{11}^2 + 2a_{12}a_{21} + a_{22}^2.$$

Consequently,

$$g^2(A) = N^2(A) - |\lambda_1(A)|^2 - |\lambda_2(A)|^2 =$$
$$a_{11}^2 + a_{12}^2 + a_{21}^2 + a_{22}^2 - (a_{11}^2 + 2a_{12}a_{21} + a_{22}^2).$$

Hence,

$$g(A) = |a_{12} - a_{21}|. \tag{2.2}$$

Example 2.2.2 *Let A be an upper-triangular matrix:*

$$A = \begin{pmatrix} a_{11} & a_{12} & \dots & a_{1n} \\ 0 & a_{21} & \dots & a_{2n} \\ . & \dots & . & . \\ 0 & \dots & 0 & a_{nn} \end{pmatrix}.$$

Then

$$g(A) = \sqrt{\sum_{k=1}^{n} \sum_{j=1}^{k-1} |a_{jk}|^2}, \qquad (2.3)$$

since the eigenvalues of a triangular matrix are its diagonal elements.

Example 2.2.3 *Consider the matrix*

$$A = \begin{pmatrix} -a_1 & \dots & -a_{n-1} & -a_n \\ 1 & \dots & 0 & 0 \\ . & \dots & . & . \\ 0 & \dots & 1 & 0 \end{pmatrix}$$

with complex numbers a_k. Such matrices play a key role in the theory of scalar ordinary differential equations. Take into account that

$$A^2 = \begin{pmatrix} a_1^2 - a_2 & \dots & a_1 a_{n-1} - a_1 & a_1 a_n \\ -a_1 & \dots & -a_{n-1} & -a_n \\ 1 & \dots & 0 & 0 \\ . & \dots & . & . \\ 0 & \dots & 0 & 0 \end{pmatrix}.$$

Thus, we obtain, $Trace\ A^2 = a_1^2 - 2a_2$. Therefore

$$g^2(A) \leq N^2(A) - |Trace\ A^2| = n - 1 - |a_1^2 - 2a_2| + \sum_{k=1}^{n} |a_k|^2. \qquad (2.4)$$

Now let a_k be real. Then (2.4) gives us the inequality

$$g^2(A) \leq n - 1 + 2a_2 + \sum_{k=2}^{n} a_k^2. \qquad (2.5)$$

2.3 Relations for Eigenvalues

Theorem 2.3.1 *For any linear operator A in $\mathbf{C}^n$,*

$$g^2(A) := N^2(A) - \sum_{k=1}^{n} |\lambda_k(A)|^2 = 2N^2(A_I) - 2\sum_{k=1}^{n} |Im\ \lambda_k(A)|^2, \qquad (3.1)$$

where $\lambda_k(A)$ are the eigenvalues of A with their multiplicities and $A_I = (A - A^*)/2i$.

To prove this theorem we need the following two lemmas.

Lemma 2.3.2 *For any linear operator A in $\mathbf{C}^n$,*

$$N^2(V) = g^2(A) \equiv N^2(A) - \sum_{k=1}^{n} |\lambda_k(A)|^2,$$

where V is the nilpotent part of A (see Section 1.7).

Proof: Let D be the diagonal part of A. Then, due to Lemma 1.7.1 both matrices V^*D and D^*V are nilpotent. Therefore,

$$Trace\ (D^*V) = 0 \text{ and } Trace\ (V^*D) = 0. \tag{3.2}$$

It is easy to see that

$$Trace\ (D^*D) = \sum_{k=1}^{n} |\lambda_k(A)|^2. \tag{3.3}$$

Since

$$A = D + V, \tag{3.4}$$

due to (3.2) and (3.3)

$$N^2(A) = Trace\ (D + V)^*(V + D) = Trace\ (V^*V + D^*D) =$$

$$N^2(V) + \sum_{k=1}^{n} |\lambda_k(A)|^2, \tag{3.5}$$

and the required equality is proved. $\square$

Lemma 2.3.3 *For any linear operator A in $\mathbf{C}^n$,*

$$N^2(V) = 2N^2(A_I) - 2\sum_{k=1}^{n} |Im\ \lambda_k(A)|^2,$$

where V is the nilpotent part of A.

Proof: Clearly,

$$-4(A_I)^2 = (A - A^*)^2 = AA - AA^* - A^*A + A^*A^*.$$

But due to (3.2) and (3.4)

$$Trace\ (A - A^*)^2 = Trace\ (V + D - V^* - D^*)^2 =$$

$$Trace\ [(V - V^*)^2 + (V - V^*)(D - D^*) +$$

$$(D - D^*)(V - V^*) + (D - D^*)^2] =$$

$$Trace \ (V - V^*)^2 + Trace \ (D - D^*)^2.$$

Hence,

$$N^2(A_I) = N^2(V_I) + N^2(D_I), \tag{3.6}$$

where

$$V_I = (V - V^*)/2i \text{ and } D_I = (D - D^*)/2i.$$

It is not hard to see that

$$N^2(V_I) = \frac{1}{2} \sum_{m=1}^{n} \sum_{k=1}^{m-1} |a_{km}|^2 = \frac{1}{2} N^2(V),$$

where a_{jk} are the entries of V in the Schur basis. Thus,

$$N^2(V) = 2N^2(A_I) - 2N^2(D_I).$$

But

$$N^2(D_I) = \sum_{k=1}^{n} |Im \ \lambda_k(A)|^2.$$

Thus, we arrive at the required equality. □

The assertion of Theorem 2.3.1 follows from Lemmas 2.3.2 and 2.3.3.

The inequality (1.3) follows from Theorem 2.3.1.

Furthermore, take into account that the nilpotent parts of the matrices $Ae^{i\tau}$ and $Ae^{i\tau} + zI$ with a real number τ and a complex one z, coincide. Hence, due to Lemma 2.3.2 we obtain the following

Corollary 2.3.4 *For any linear operator A in $\mathbf{C}^n$, a real number τ, and a complex one z, relation (1.4) holds.*

Corollary 2.3.5 *For arbitrary commuting linear operators A, B in $\mathbf{C}^n$,*

$$g(A + B) \le g(A) + g(B).$$

In fact, A and B have the same Schur basis. This clearly forces

$$V_{A+B} = V_A + V_B,$$

where V_{A+B}, V_A and V_B are the nilpotent parts of $A + B, A$ and B, respectively. Due to Lemma 2.3.2 the relations

$$g(A) = N(V_A), \ g(B) = N(V_B), \ g(A + B) = N(V_{A+B})$$

are true. Now the property of the norm implies the result.

Corollary 2.3.6 *For any $n \times n$ matrix A and real numbers t, τ the equality*

$$N^2(Ae^{it} - A^*e^{-it}) - \sum_{k=1}^{n} |e^{it}\lambda_k(A) - e^{-it}\overline{\lambda}_k(A)|^2 =$$

$$N^2(Ae^{i\tau} - A^*e^{-i\tau}) - \sum_{k=1}^{n} |e^{i\tau}\lambda_k(A) - e^{-i\tau}\overline{\lambda}_k(A)|^2$$

is true.

The proof consists in replacing A by Ae^{it} and $Ae^{i\tau}$ and using Theorem 2.3.1. In particular, take $t = 0$ and $\tau = \pi/2$. Due to Corollary 2.3.6,

$$N^2(A_I) - \sum_{k=1}^{n} |Im\ \lambda_k(A)|^2 = N^2(A_R) - \sum_{k=1}^{n} |Re\ \lambda_k(A)|^2$$

with $A_R = (A + A^*)/2$.

2.4   An Auxiliary Inequality

Lemma 2.4.1 *For arbitrary positive numbers $a_1, ..., a_n$ and $m = 1, ..., n$, we have*

$$\sum_{1 \le k_1 < k_2 < ... < k_m \le n} a_{k_1} ... a_{k_m} \le n^{-m} C_n^m \ [\sum_{k=1}^{n} a_k]^m. \qquad (4.1)$$

Proof: Consider the following function of n positive variables $y_1, ..., y_n$:

$$R_m(y_1, ..., y_n) \equiv \sum_{1 \le k_1 < k_2 < ... < k_m \le n} y_{k_1} y_{k_2} \cdots y_{k_m}.$$

Let us prove that under the condition

$$\sum_{k=1}^{n} y_k = n\, b \qquad (4.2)$$

where b is a given positive number, function R_m has a unique conditional maximum. To this end denote

$$F_j(y_1, ..., y_n) \equiv \frac{\partial R_m(y_1, ..., y_n)}{\partial y_j}.$$

Obviously, $F_j(y_1, ..., y_n)$ does not depend on y_j, symmetrically depends on other variables, and monotonically increases with respect to each of its variables. The conditional extremums of R_m under (4.2) are the roots of the equations

$$F_j(y_1, ..., y_n) - \lambda \frac{\partial}{\partial y_j} \sum_{k=1}^{n} y_k = 0 \ (j = 1, ..., n),$$

where λ is the Lagrange factor. Therefore,

$$F_j(y_1, ..., y_n) = \lambda \; (j = 1, ..., n).$$

Since $F_j(y_1, ..., y_n)$ does not depend on y_j, and $F_k(y_1, ..., y_n)$ does not depend on y_k, equality

$$F_j(y_1, ..., y_n) = F_k(y_1, ..., y_n) = \lambda$$

for all $k \neq j$ is possible if and only if $y_j = y_k$. Thus R_m has under (4.2) a unique extremum when

$$y_1 = y_2 = ... = y_n = b. \tag{4.3}$$

But

$$R_m(b, ..., b) = b^m \sum_{1 \le k_1 < k_2 < ... < k_m \le n} 1 = b^m C_n^m. \tag{4.4}$$

Let us check that (4.3) gives us the maximum. Letting

$$y_1 \to nb \text{ and } y_k \to 0 \; (k = 2, ..., n),$$

we get

$$R_m(y_1, ..., y_n) \to 0.$$

Since the extremum (4.3) is unique, it is the maximum. Thus, under (4.2)

$$R_m(y_1, ..., y_n) \le b^m C_n^m \; (y_k \ge 0, \; k = 1, ..., n).$$

Take $y_j = a_j$ and

$$b = \frac{a_1 + ... + a_n}{n}.$$

Then

$$R_m(a_1, ..., a_n) \le C_n^m n^{-m} [\sum_{k=1}^{n} a_k]^m.$$

We thus get the required result. $\square$

2.5 Euclidean Norms of Powers of Nilpotent Matrices

Theorem 2.5.1 *For any nilpotent operator V in $\mathbf{C}^n$, the inequalities*

$$\|V^k\| \le \gamma_{n,k} N^k(V) \; (k = 1, ..., n-1)$$

are valid.

Proof: Since V is nilpotent, due to the Shur theorem we can represent it by an upper-triangular matrix with the zero diagonal:

$$V = (a_{jk})_{j,k=1}^n \text{ with } a_{jk} = 0 \ (j \geq k).$$

Denote

$$\|x\|_m = (\sum_{k=m}^n |x_k|^2)^{1/2} \text{ for } m < n,$$

where x_k are coordinates of a vector x. We can write

$$\|Vx\|_m^2 = \sum_{j=m}^{n-1} |\sum_{k=j+1}^n a_{jk} x_k|^2 \text{ for all } m \leq n - 1.$$

Now we have (by Schwarz's inequality) the relation

$$\|Vx\|_m^2 \leq \sum_{j=m}^{n-1} h_j \|x\|_{j+1}^2, \tag{5.1}$$

where

$$h_j = \sum_{k=j+1}^n |a_{jk}|^2 \ (j < n).$$

Further, by Schwarz's inequality

$$\|V^2 x\|_m^2 = \sum_{j=m}^{n-1} |\sum_{k=j+1}^n a_{jk} (Vx)_k|^2 \leq \sum_{j=m}^{n-1} h_j \|Vx\|_{j+1}^2.$$

Here $(Vx)_k$ are coordinates of Vx. Taking into account (5.1), we obtain

$$\|V^2 x\|_m^2 \leq \sum_{j=m}^{n-1} h_j \sum_{k=j+1}^{n-1} h_k \|x\|_{k+1}^2 = \sum_{m \leq j < k \leq n-1} h_j h_k \|x\|_{k+1}^2.$$

Hence,

$$\|V^2\|^2 \leq \sum_{1 \leq j < k \leq n-1} h_j h_k.$$

Repeating these arguments, we arrive at the inequality

$$\|V^p\|^2 \leq \sum_{1 \leq k_1 < k_2 < \ldots < k_p \leq n-1} h_{k_1} \ldots h_{k_p}. \tag{5.2}$$

Therefore due to Lemma 2.4.1,

$$\|V^k\|^2 \leq \gamma_{n,k}^2 (\sum_{j=1}^{n-1} h_j)^k.$$

But

$$\sum_{j=1}^{n-1} h_j \leq N^2(V).$$

We thus have derived the required result. □

Theorem 2.5.1 and (1.5) imply

Corollary 2.5.2 *For any nilpotent operator V in $\mathbf{C}^n$, the inequalities*

$$\|V^k\| \leq \frac{N^k(V)}{\sqrt{k!}} \quad (k = 1, \ldots, n-1)$$

are valid.

An independent proof of this corollary can be found in (Gil', 1995, p. 50, Lemma 2.3.1).

2.6   Proof of Theorem 2.1.1

Let D and V be the diagonal and nilpotent parts of A, respectively. According to Lemma 1.7.1, $R_\lambda(D)V$ is a nilpotent operator. So by virtue of Theorem 2.5.1,

$$\|(R_\lambda(D)V)^k\| \leq N^k(R_\lambda(D)V)\gamma_{n,k} \quad (k = 1, \ldots, n-1). \tag{6.1}$$

Since D is a normal operator, we can write down $\|R_\lambda(D)\| = \rho^{-1}(D, \lambda)$. It is clear that

$$N(R_\lambda(D)V) \leq N(V)\|R_\lambda(D)\| = N(V)\rho^{-1}(D, \lambda). \tag{6.2}$$

According to (3.4),

$$A - \lambda I = D + V - \lambda I = (D - \lambda I)(I + R_\lambda(D)V).$$

We thus have

$$R_\lambda(A) = (I + R_\lambda(D)V)^{-1}R_\lambda(D) = \sum_{k=0}^{n-1}(R_\lambda(D)V)^k R_\lambda(D). \tag{6.3}$$

Now (6.1) and (6.2) yield the inequality

$$\|R_\lambda(A)\| \leq \sum_{k=0}^{n-1} N^k(V)\gamma_{n,k}\rho^{-k-1}(D, \lambda).$$

This relation proves the stated result, since A and D have the same eigenvalues and $N(V) = g(A)$, due to Lemma 2.3.2. □

An additional proof of Corollary 2.1.2:   Corollary 2.5.2 implies

$$\|(R_\lambda(D)V)^k\| \leq \frac{N^k(R_\lambda(D)V)}{\sqrt{k!}} \quad (k = 1, \ldots, n-1).$$

Now the required result follows from (6.2) and (6.3), since $N(V) = g(A)$ due to Lemma 2.3.2. □

2.7 Estimates for the Norm of Analytic Matrix-Valued Functions

Recall that $g(A)$ and $\gamma_{n,k}$ are defined in Section 2.1, and $B(\mathbf{C}^n)$ is the set of all linear operators in $\mathbf{C}^n$.

Theorem 2.7.1 *Let $A \in B(\mathbf{C}^n)$ and let f be a function regular on a neighborhood of the closed convex hull $co(A)$ of the eigenvalues of A. Then*

$$\|f(A)\| \le \sum_{k=0}^{n-1} \sup_{\lambda \in co(A)} |f^{(k)}(\lambda)| g^k(A) \frac{\gamma_{n,k}}{k!}. \tag{7.1}$$

The proof of this theorem is divided into a series of lemmas, which are presented in the next section.

Theorem 2.7.1 is exact: if A is a normal matrix and

$$\sup_{\lambda \in co(A)} |f(\lambda)| = \sup_{\lambda \in \sigma(A)} |f(\lambda)|,$$

then we have the equality $\|f(A)\| = \sup_{\lambda \in \sigma(A)} |f(\lambda)|$. Theorem 2.7.1 and inequalities (1.5) yield

Corollary 2.7.2 *Let $A \in B(\mathbf{C}^n)$ and let f be a function regular on a neighborhood of the closed convex hull $co(A)$ of the eigenvalues of A. Then*

$$\|f(A)\| \le \sum_{k=0}^{n-1} \sup_{\lambda \in co(A)} |f^{(k)}(\lambda)| \frac{g^k(A)}{(k!)^{3/2}}. \tag{7.2}$$

An additional proof of this corollary is presented in the next section.

Example 2.7.3 *For a linear operator A in $\mathbf{C}^n$, Theorem 2.7.1 and Corollary 2.7.2 give us the estimates*

$$\|exp(At)\| \le e^{\alpha(A)t} \sum_{k=0}^{n-1} g^k(A) t^k \frac{\gamma_{n,k}}{k!} \le e^{\alpha(A)t} \sum_{k=0}^{n-1} \frac{g^k(A) t^k}{(k!)^{3/2}} \quad (t \ge 0)$$

where $\alpha(A) = \max_{k=1,\dots,n} Re\, \lambda_k(A)$. In addition,

$$\|A^m\| \le \sum_{k=0}^{n-1} \frac{\gamma_{n,k} m! g^k(A) r_s^{m-k}(A)}{(m-k)! k!} \le \sum_{k=0}^{n-1} \frac{m! g^k(A) r_s^{m-k}(A)}{(m-k)!(k!)^{3/2}} \quad (m = 1, 2, \dots)$$

where $r_s(A)$ is the spectral radius. Recall that $1/(m-k)! = 0$ if $m < k$.

2.8   Proof of Theorem 2.7.1

Lemma 2.8.1 *Let $\{d_k\}$ be an orthogonal normal basis in $\mathbf{C}^n$, $A_1, \ldots, A_j$ $n \times n$-matrices and $\phi(k_1, \ldots, k_{j+1})$ a scalar-valued function of arguments*

$$k_1, \ldots, k_{j+1} = 1, 2, \ldots, n; \ j < n.$$

Define projectors $Q(k)$ by $Q(k)h = (h, d_k)d_k$ $(h \in \mathbf{C}^n, \ k = 1, \ldots, n)$, and set

$$T = \sum_{1 \leq k_1, \ldots, k_{j+1} \leq n} \phi(k_1, \ldots, k_{j+1}) Q(k_1) A_1 Q(k_2) \ldots A_j Q(k_{j+1}).$$

Then $\|T\| \leq a(\phi)\| |A_1||A_2| \ldots |A_j| \|$, where

$$a(\phi) = \max_{1 \leq k_1, \ldots, k_{j+1} \leq n} |\phi(k_1, \ldots, k_{j+1})|$$

and $|A_k|$ $(k = 1, \ldots, j)$ are the matrices, whose entries in $\{d_k\}$ are the absolute values of the entries of A_k in $\{d_k\}$.

Proof: For any entry $T_{sm} = (Td_s, d_m)$ $(s, m = 1, \ldots, n)$ of operator T we have

$$T_{sm} = \sum_{1 \leq k_2, \ldots, k_j \leq n} \phi(s, k_2, \ldots, k_j, m) a_{sk_2}^{(1)} \ldots a_{k_j, m}^{(j)},$$

where $a_{jk}^{(i)} = (A_i d_k, d_j)$ are the entries of A_i. Hence,

$$|T_{sm}| \leq a(\phi) \sum_{1 \leq k_2, \ldots, k_j \leq n} |a_{sk_2}^{(1)} \ldots a_{k_j m}^{(j)}|.$$

This relation and the equality

$$\|Tx\|^2 = \sum_{j=1}^{n} |(Tx)_j|^2 \ (x \in \mathbf{C}^n),$$

where $(.)_j$ means the j-th coordinate, imply the required result. $\square$

Furthermore, let $|V|$ be the operator whose matrix elements in the orthonormed basis of the triangular representation (the Schur basis) are the absolute values of the matrix elements of the nilpotent part V of A with respect to this basis. That is,

$$|V| = \sum_{k=1}^{n} \sum_{j=1}^{k-1} |a_{jk}|(., e_k)e_j,$$

where $\{e_k\}$ is the Schur basis and $a_{jk} = (Ae_k, e_j)$.

Lemma 2.8.2 *Under the hypothesis of Theorem 2.7.1, the estimate*

$$\|f(A)\| \leq \sum_{k=0}^{n-1} \sup_{\lambda \in co(A)} |f^{(k)}(\lambda)| \frac{\| |V|^k \|}{k!}$$

is true, where V is the nilpotent part of A.

Proof: It is not hard to see that the representation (3.4) implies the equality

$$R_\lambda(A) \equiv (A - I\lambda)^{-1} = (D + V - \lambda I)^{-1} =$$

$$(I + R_\lambda(D)V)^{-1} R_\lambda(D)$$

for all regular λ. According to Lemma 1.7.1 $R_\lambda(D)V$ is a nilpotent operator because V and $R_\lambda(D)$ have common invariant subspaces. Hence,

$$(R_\lambda(D)V)^n = 0.$$

Therefore,

$$R_\lambda(A) = \sum_{k=0}^{n-1} (R_\lambda(D)V)^k (-1)^k R_\lambda(D). \tag{8.1}$$

Due to the representation for functions of matrices

$$f(A) = -\frac{1}{2\pi i} \int_\Gamma f(\lambda) R_\lambda(A) d\lambda = \sum_{k=0}^{n-1} C_k, \tag{8.2}$$

where

$$C_k = (-1)^{k+1} \frac{1}{2\pi i} \int_\Gamma f(\lambda)(R_\lambda(D)V)^k R_\lambda(D) d\lambda.$$

Here Γ is a closed contour surrounding $\sigma(A)$. Since D is a diagonal matrix with respect to the Schur basis $\{e_k\}$ and its diagonal entries are the eigenvalues of A, then

$$R_\lambda(D) = \sum_{j=1}^{n} \frac{Q_j}{\lambda_j(A) - \lambda},$$

where $Q_k = (., e_k)e_k$. We have

$$C_k = \sum_{j_1=1}^{n} Q_{j_1} V \sum_{j_2=1}^{n} Q_{j_2} V \ldots V \sum_{j_k=1}^{n} Q_{j_k} I_{j_1 j_2 \ldots j_{k+1}}.$$

Here

$$I_{j_1 \ldots j_{k+1}} = \frac{(-1)^{k+1}}{2\pi i} \int_\Gamma \frac{f(\lambda) d\lambda}{(\lambda_{j_1} - \lambda) \ldots (\lambda_{j_{k+1}} - \lambda)}.$$

Lemma 2.8.1 gives us the estimate

$$\|C_k\| \leq \max_{1 \leq j_1 \leq \cdots \leq j_{k+1} \leq n} |I_{j_1 \ldots j_{k+1}}| \|\, |V|^k \,\|.$$

Due to Lemma 1.5.1

$$|I_{j_1 \ldots j_{k+1}}| \leq \sup_{\lambda \in co(A)} \frac{|f^{(k)}(\lambda)|}{k!}.$$

This inequality and (8.2) imply the result. $\square$

Proof of Theorem 2.7.1: Theorem 2.5.1 implies

$$\|\, |V|^k \,\| \leq \gamma_{n,k} N^k(|V|) \; (k = 1, ..., n-1).$$

But $N(|V|) = N(V)$. Moreover, thanks to Lemma 2.3.2, $N(V) = g(A)$. Thus

$$\|\, |V|^k \,\| \leq \gamma_{n,k} g^k(A) \; (k = 1, ..., n-1).$$

Now the previous lemma yields the required result. $\square$

An additional proof of Corollary 2.7.2: Corollary 2.5.2 implies

$$\|\, |V|^k \,\| \leq \frac{N^k(V)}{\sqrt{k!}} \; (k = 1, ..., n-1)$$

Now the required result follows from Lemma 2.8.2, since $N(V) = g(A)$ due to Lemma 2.3.2. $\square$

2.9 The First Multiplicative Representation of the Resolvent

Recall that $B(\mathbf{C}^n)$ is the set of linear operators in $\mathbf{C}^n$, I is the unit operator. Let $P_k \; (k = 1, \ldots, n)$ be *the maximal chain of the invariant projectors of an* $A \in B(\mathbf{C}^n)$. That is, P_k are orthogonal projectors,

$$AP_k = P_k AP_k \; (k = 1, \ldots, n)$$

and

$$0 = P_0 \mathbf{C}^n \subset P_1 \mathbf{C}^n \subset \ldots \subset P_n \mathbf{C}^n = \mathbf{C}^n.$$

So $dim \, \Delta P_k = 1$. Here

$$\Delta P_k = P_k - P_{k-1} \; (k = 1, ..., n).$$

We use the triangular representation

$$A = D + V. \tag{9.1}$$

(see Section 1.7). Here V is the nilpotent part of A and

$$D = \sum_{k=1}^{n} \lambda_k(A)\Delta P_k$$

is the diagonal part. For $X_1, X_2, ..., X_n \in B(\mathbf{C}^n)$ denote

$$\overrightarrow{\prod_{1\leq k\leq n}} X_k \equiv X_1 X_2 ... X_n.$$

That is, the arrow over the symbol of the product means that the indexes of the co-factors increase from left to right.

Theorem 2.9.1 *For any $A \in B(\mathbf{C}^n)$,*

$$\lambda R_\lambda(A) = - \overrightarrow{\prod_{1\leq k\leq n}} (I + \frac{A\Delta P_k}{\lambda - \lambda_k(A)}) \ (\lambda \notin \sigma(A)),$$

where P_k, $k = 1, ..., n$ is the maximal chain of the invariant projectors of A.

Proof: Denote $E_k = I - P_k$. Since

$$A = (E_k + P_k)A(E_k + P_k) \text{ for any } k = 1, ..., n$$

and $E_1 A P_1 = 0$, we get the relation

$$A = P_1 A E_1 + P_1 A P_1 + E_1 A E_1.$$

Take into account that $\Delta P_1 = P_1$ and

$$P_1 A P_1 = \lambda_1(A)\Delta P_1.$$

Then

$$A = \lambda_1(A)\Delta P_1 + P_1 A E_1 + E_1 A E_1 =$$
$$\lambda_1(A)\Delta P_1 + A E_1. \tag{9.2}$$

Now, we check the equality

$$R_\lambda(A) = \Psi(\lambda), \tag{9.3}$$

where

$$\Psi(\lambda) \equiv \frac{\Delta P_1}{\lambda_1(A) - \lambda} - \frac{\Delta P_1}{\lambda_1(A) - \lambda} A E_1 R_\lambda(A) E_1 + E_1 R_\lambda(A) E_1.$$

In fact, multiplying this equality from the left by $A - I\lambda$ and taking into account equality (9.2), we obtain the relation

$$(A - I\lambda)\Psi(\lambda) = \Delta P_1 - \Delta P_1 A E_1 R_\lambda(A) E_1 + (A - I\lambda)E_1 R_\lambda(A) E_1.$$

But $E_1 A E_1 = E_1 A$ and thus $E_1 R_\lambda(A) E_1 = E_1 R_\lambda(A)$. I.e. we can write

$$(A - I\lambda)\Psi(\lambda) = \Delta P_1 + (-\Delta P_1 A + A - I\lambda) E_1 R_\lambda(A) =$$

$$\Delta P_1 + E_1(A - I\lambda) R_\lambda(A) = \Delta P_1 + E_1 = I.$$

Similarly, we multiply (9.3) by $A - I\lambda$ from the right and take into account (9.2). This gives I. Therefore, (9.3) is correct.

Due to (9.3)

$$I - A R_\lambda(A) =$$

$$(I - (\lambda_1(A) - \lambda)^{-1} A \Delta P_1)(I - A E_1 R_\lambda(A) E_1). \tag{9.4}$$

Now we apply the above arguments to operator $A E_1$. We obtain the following expression which is similar to (9.4):

$$I - A E_1 R_\lambda(A) E_1 =$$

$$(I - (\lambda_2(A) - \lambda)^{-1} A \Delta P_2)(I - A E_2 R_\lambda(A) E_2).$$

For any $k < n$, it similarly follows that

$$I - A E_k R_\lambda(A) E_k =$$

$$(I - \frac{A \Delta P_{k+1}}{\lambda_{k+1}(A) - \lambda})(I - A E_{k+1} R_\lambda(A) E_{k+1}).$$

Substitute this in (9.4), as long as $k = 1, 2, ..., n - 1$. We have

$$I - A R_\lambda(A) =$$

$$\overset{\rightarrow}{\prod_{1 \le k \le n-1}} (I + \frac{A \Delta P_k}{\lambda - \lambda_k(A)})(I - A E_{n-1} R_\lambda(A) E_{n-1}). \tag{9.5}$$

It is clear that $E_{n-1} = \Delta P_n$. I.e.,

$$I - A E_{n-1} R_\lambda(A) E_{n-1} = I + \frac{A \Delta P_n}{\lambda - \lambda_n(A)}.$$

Now the identity

$$I - A R_\lambda(A) = -\lambda R_\lambda(A)$$

and (9.5) imply the result. $\square$

Let A be a normal matrix. Then

$$A = \sum_{k=1}^{n} \lambda_k(A) \Delta P_k.$$

Hence, $A \Delta P_k = \lambda_k(A) \Delta P_k$. Since $\Delta P_k \Delta P_j = 0$ for $j \ne k$, Theorem 2.9.1 gives us the equality

$$\lambda R_\lambda(A) = -\sum_{k=1}^{n}(I + (\lambda - \lambda_k(A))^{-1} \lambda_k(A) \Delta P_k).$$

But

$$I = \sum_{k=1}^{n} \Delta P_k.$$

The result is

$$\lambda R_\lambda(A) = -\sum_{k=1}^{n} [1 + (\lambda - \lambda_k(A))^{-1}\lambda_k(A)]\Delta P_k) =$$

$$-\sum_{k=1}^{n} \lambda \frac{\Delta P_k}{\lambda - \lambda_k(A)}.$$

Or

$$R_\lambda(A) = \sum_{k=1}^{n} \frac{\Delta P_k}{\lambda_k(A) - \lambda}.$$

We have obtained the well-known spectral representation for the resolvent of a normal matrix.

Thus, Theorem 2.9.1 generalizes the spectral representation for the resolvent of a normal matrix.

2.10 The Second Multiplicative Representation of the Resolvent

Lemma 2.10.1 *Let $V \in B(\mathbf{C}^n)$ be a nilpotent operator and P_k, $k = 1, ..., n$, be the maximal chain of its invariant projectors. Then*

$$(I - V)^{-1} = \prod_{2 \le k \le n}^{\rightarrow} (I + V\Delta P_k). \tag{10.1}$$

Proof: In fact, all the eigenvalues of V are equal to zero, and $V\Delta P_1 = 0$. Now Theorem 2.9.1 gives us relation (10.1). $\square$

Relation (10.1) allows us to prove the second multiplicative representation of the resolvent of A.

Theorem 2.10.2 *Let D and V be the diagonal and nilpotent parts of $A \in B(\mathbf{C}^n)$, respectively. Then*

$$R_\lambda(A) = R_\lambda(D) \prod_{2 \le k \le n}^{\rightarrow} [I + \frac{V\Delta P_k}{\lambda - \lambda_k(A)}] \ (\lambda \notin \sigma(A)), \tag{10.2}$$

where P_k, $k = 1, ..., n$, is the maximal chain of invariant projectors of A.

Proof: Due to (9.1)

$$R_\lambda(A) = (A - \lambda I)^{-1} = (D + V - \lambda I)^{-1} = R_\lambda(D)(I + VR_\lambda(D))^{-1}.$$

But $VR_\lambda(D)$ is a nilpotent operator. Take into account that

$$R_\lambda(D)\Delta P_k = (\lambda_k(A) - \lambda)^{-1}\Delta P_k.$$

Now (10.1) ensures the relation (10.2). $\square$

2.11   The First Relation between Determinants and Resolvents

In this section, a relation between the determinant and resolvent of a matrix is derived. It improves the Carleman inequality. We recall the Carleman inequality in Section 2.15.

Theorem 2.11.1 *Let $A \in B(\mathbf{C}^n)$ $(n > 1)$ and $I - A$ be invertible. Then*

$$\|(I - A)^{-1} \det (I - A)\| \leq$$

$$[1 + \frac{1}{n-1} (N^2(A) - 2Re\, Trace\,(A) + 1)]^{(n-1)/2}. \qquad (11.1)$$

The proof of this theorem is presented in this section below.

Corollary 2.11.2 *Let $A \in \mathbf{B}(\mathbf{C}^n)$. Then*

$$\|(I\lambda - A)^{-1} \det (\lambda I - A)\| \leq$$

$$[\frac{N^2(A) - 2Re\, (\overline{\lambda}\, Trace\,(A))\, + n|\lambda|^2}{n-1}]^{(n-1)/2} \quad (\lambda \notin \sigma(A)).$$

$$(11.2)$$

In particular, let V be a nilpotent matrix. Then

$$\|(I\lambda - V)^{-1}\| \leq$$

$$\frac{1}{|\lambda|} [1 + \frac{1}{n-1} (1 + \frac{N^2(V)}{|\lambda|^2})]^{(n-1)/2} \quad (\lambda \neq 0). \qquad (11.3)$$

Indeed, inequality (11.2) is due to Theorem 11.1 with $\lambda^{-1}A$ instead of A, and the equality $|\lambda|^2\lambda^{-1} = \overline{\lambda}$ taken into account. If V is nilpotent, then

$$|\det (\lambda I - V)|^2 = \det (I\lambda - V) \det (I\overline{\lambda} - V^*) = |\lambda|^{2n}.$$

Moreover, $Trace\, V = 0$. So (11.2) implies (11.3).

   To prove Theorem 2.11.1, we need the following

Lemma 2.11.3 *Let* $A \in B(\mathbf{C}^n)$ *be a positive definite Hermitian matrix:* $A = A^* > 0$. *Then*

$$\|A^{-1} \det A\| \leq [\frac{Trace\ A}{n-1}]^{n-1}.$$

Proof: Without loss of generality assume that

$$\lambda_n(A) = \min_{k=1,\ldots,n} \lambda_k(A).$$

Then $\|A^{-1}\| = \lambda_n^{-1}(A)$ and

$$\|A^{-1} \det A\| = \prod_{k=1}^{n-1} \lambda_k(A).$$

Hence, due to the inequality between the arithmetic and geometric mean values we get

$$\|A^{-1} \det A\| \leq [(n-1)^{-1} \sum_{k=1}^{n-1} \lambda_k]^{n-1} \leq [(n-1)^{-1} Trace\ A]^{n-1},$$

since A is positive definite. As claimed. $\square$

Proof of Theorem 2.11.1: For any $A \in B(\mathbf{C}^n)$, the operator

$$B \equiv (I - A)(I - A^*)$$

is positive definite and

$$\det B = \det\ (I - A)(I - A^*) = \det\ (I - A) \det\ (I - A^*) =$$
$$|\det\ (I - A)|^2.$$

Moreover,

$$Trace\ [(I - A)(I - A^*)] = Trace\ I - Trace\ (A + A^*) + Trace\ (AA^*) =$$
$$n - 2Re\ Trace\ A + N^2(A).$$

But

$$\|B^{-1}\| = \|(I - A)^{-1}(I - A^*)^{-1}\| = \|(I - A)^{-1}\|^2.$$

Now Lemma 2.11.2 yields

$$\|B^{-1} \det B\| = \|(I - A)^{-1} \det\ (I - A)\|^2 \leq$$

$$[\frac{1}{n-1}\ (n + N^2(A) - 2Re\ Trace\ (A))]^{n-1} = .$$

$$[1 + \frac{1}{n-1}\ (N^2(A) - 2Re\ Trace\ (A) + 1)]^{n-1},$$

as claimed. $\square$

2.12 The Second Relation between Determinants and Resolvents

Without any loss of generality assume that for a regular λ the relation

$$|\lambda - \lambda_1(A)| = min_{k=1,...,n}|\lambda - \lambda_k(A)| \tag{12.1}$$

is valid. Recall that $g(A)$ and $\gamma_{n,k}$ are defined in Section 2.1.

Theorem 2.12.1 *Let $A \in B(\mathbf{C}^n)$ and (12.1) hold. Then*

$$\|R_\lambda(A)det(A - I\lambda)\| \le \prod_{j=2}^{n} max\,\{1, |\lambda_j(A) - \lambda|\}G(A), \tag{12.2}$$

where

$$G(A) = \sum_{k=0}^{n-1} g^k(A)\gamma_{n,k}.$$

The proof of this theorem is given in the next section.

Theorem 2.12.1 is exact. In fact, for instance, let A be a unitary operator and $\lambda = 0$. Since any unitary operator is normal and $|\lambda_k(A)| = 1$, then $G(A) = 1$ and due to (12.3)

$$|detA| = 1 \le \|A\|.$$

But the norm of a unitary operator equals 1. Thus, we arrive at the equality

$$|detA| = \|A\| = 1.$$

Let $r_s(A)$ be the spectral radius of a matrix A. Let A be nonsingular. Put $A_0 = r^{-1}(A)A$. Due to Theorem 2.12.1,

$$\|A_0^{-1}det(A_0)\| \le G(A_0).$$

But $g(A_0) = r_s^{-1}(A)\,g(A)$. Therefore, the following result holds

Corollary 2.12.2 *Let $A \in B(\mathbf{C}^n)$ be nonsingular. Then*

$$\|A^{-1}det(A)\| \le \sum_{k=0}^{n-1} r_s^{n-k-1}(A)g^k(A)\gamma_{n,k}.$$

2.13 Proof of Theorem 2.12.1

For brevity, put

$$\lambda_j(A) = \lambda_j \ (j = 1, ..., n).$$

Without any loss of generality assume that

$$|\lambda_1| = \min_{k=1,...,n} |\lambda_k| \tag{13.1}$$

Lemma 2.13.1 *Let A be a nonsingular operator in $\mathbf{C}^n$ and condition (13.1) hold. Then with the notations*

$$\chi_k = \max\{1, |\lambda_k|\}$$

and

$$\omega(A) = \prod_{j=2}^{n} \chi_j,$$

the inequality

$$\|\det(A)A^{-1}\| \leq G(A).$$

holds.

Proof: We have

$$\|A^{-1}\| = \|D^{-1}(I + VD^{-1})^{-1}\| \leq \|(I + VD^{-1})^{-1}\|\, \|D^{-1}\|. \qquad (13.2)$$

Clearly, $\|D^{-1}\| = |\lambda_1|^{-1}$. To estimate $\|(I + VD^{-1})^{-1}\|$, observe that VD^{-1} is a nilpotent matrix and, due to Lemma 2.10.1,

$$(I + VD^{-1})^{-1} = \overset{\rightarrow}{\prod_{2 \leq k \leq n}} (I - \lambda_k^{-1} V \Delta P_k),$$

since

$$D^{-1} \Delta P_k = \lambda_k^{-1} \Delta P_k.$$

This yields

$$(I + VD^{-1})^{-1} = \prod_{k=2}^{n} \lambda_k^{-1}\, K = \lambda_1 [\det(A)]^{-1}\, K, \qquad (13.3)$$

where

$$K = \overset{\rightarrow}{\prod_{2 \leq k \leq n}} (I \lambda_k - V \Delta P_k).$$

It not hard to check that

$$|I \lambda_k - V \Delta P_k| \leq (I + |V| \Delta P_k)\, \chi_k.$$

Here $|A|$ means the matrix $(\,|a_{ij}|\,)$, if $A = (a_{ij})$. The inequality $B \leq C$ for non-negative matrices B, C is understood in the natural sense. So we have

$$|K| \leq \overset{\rightarrow}{\prod_{2 \leq k \leq n}} \chi_k\, (I + |V| \Delta P_k) = \omega(A) \overset{\rightarrow}{\prod_{2 \leq k \leq n}} (I + |V| \Delta P_k) =$$

$$\omega(A)\, (I - |V|)^{-1}.$$

Taking into account that $N(V) = N(|V|)$ and $N(V) = g(A)$, by virtue of Theorem 2.1.1 we get the inequality

$$\|(I - |V|)^{-1}\| \le \sum_{k=0}^{n} \gamma_{k,n} N^k(V) = G(A).$$

That is,

$$\|K\| \le \omega(A)G(A).$$

Due to (13.2) and (13.3)

$$\|A^{-1}\| \le |det(A)|^{-1}\omega(A)G(A).$$

As claimed. □

The assertion of Theorem 2.12.1 follows from the latter lemma with $A - \lambda I$ instead of A. □

2.14   An Additional Estimate for Resolvents

Theorem 2.14.1 *Let $A \in \mathbf{B}(\mathbf{C}^n)$, $n > 1$. Then*

$$\|(I\lambda - A)^{-1}\| \le \frac{1}{\rho(A, \lambda)}[1 + \frac{1}{n-1}(1 + \frac{g^2(A)}{\rho^2(A, \lambda)})]^{(n-1)/2}$$

for any regular λ of A.

Proof: Due to the triangular representation (see Section 1.7),

$$(I\lambda - A)^{-1} = (I\lambda - D - V)^{-1} = (I\lambda - D)^{-1}(I + B_\lambda)^{-1}, \qquad (14.1)$$

where $B_\lambda := -V(I\lambda - D)^{-1}$. But operator B_λ is a nilpotent one. So Theorem 2.11.1 implies

$$\|(I + B_\lambda)^{-1}\| \le [1 + \frac{1 + N^2(B_\lambda)}{n-1}]^{(n-1)/2}. \qquad (14.2)$$

Take into account that

$$N(B_\lambda) = N(V(I\lambda - D)^{-1}) \le \|(I\lambda - D)^{-1}\|N(V) = \rho^{-1}(D, \lambda)N(V).$$

Moreover, $\sigma(D)$ and $\sigma(A)$ coincide and due to Lemma 2.3.2, $N(V) = g(A)$. Thus,

$$N(B_\lambda) \le \rho^{-1}(D, \lambda)g(A) = \rho^{-1}(A, \lambda)g(A).$$

Now (14.1) and (14.2) imply the required result. □

2.15   Notes

The quantity $g(A)$ was introduced both by P. Henrici (1962) and independently by M.I. Gil' (1979b).

Theorem 2.1.1 was derived in the paper (Gil', 1979a) in a more general situation and was extended in (Gil', 1995) (see also (Gil', 1993b)).

Recall that Carleman has derived the inequality

$$\|R_\lambda(A) \prod_{k=1}^{n} (1 - \lambda^{-1}\lambda_k(A))exp[\lambda^{-1}\lambda_k(A)]\| \le$$

$$|\lambda|exp[1 + N^2(A\lambda^{-1})/2],$$

cf. (Dunford, N and Schwartz, 1963, p. 1023).

Theorem 2.3.1 was published in (Gil', 1993a). It improves Schur's inequality

$$\sum_{k=1}^{n} |\lambda_k(A)|^2 \le N^2(A)$$

and Brown's inequality

$$\sum_{k=1}^{n} |Im\ \lambda_k(A)|^2 \le N^2(A_I)$$

(see (Marcus and Minc, 1964)). A very interesting inequality for eigenvalues of matrices was derived in (Kress et. al, 1974).

Gel'fand and G.E. Shilov (1958) have established the estimate

$$\|f(A)\| \le \sum_{k=0}^{n-1} \sup_{\lambda \in co(A)} |f^{(k)}(\lambda)|(2\|A\|)^k.$$

About other estimations for the matrix exponent, see (Coppel, 1978).

Theorem 2.7.1 was derived in the paper (Gil', 1979b) in the case of the Hilbert-Schmidt operators (see also (Gil', 1993b)).

Theorems 2.9.1 and 2.10.1 were published in the more general situation in (Gil', 1973).

Theorems 2.11.1 and 2.12.1 are probably new.

References

[1] Coppel, W.A. (1978). *Dichotomies in Stability Theory*. Lecture Notes in Mathematics, No. 629, Springer-Verlag, New York.

[2] Dunford, N and Schwartz, J. T. (1963). *Linear Operators, part II. Spectral Theory*. Interscience publishers, New York, London.

[3] Gel'fand, I.M. and Shilov, G.E. (1958). *Some Questions of Theory of Differential Equations*. Nauka, Moscow (in Russian).

[4] Gil', M. I. (1973). On the representation of the resolvent of a nonselfadjoint operator by the integral with respect to a spectral function, *Soviet Math. Dokl.*, **14** , 1214-1217.

[5] Gil', M. I. (1979a). An estimate for norms of resolvent of completely continuous operators, *Mathematical Notes*, **26** , 849-851.

[6] Gil', M. I. (1979b). Estimating norms of functions of a Hilbert-Schmidt operator (in Russian), *Izvestiya VUZ, Matematika*, **23**, 14-19. English translation in *Soviet Math.*, **23** , 13-19.

[7] Gil', M. I. (1983). One estimate for resolvents of nonselfadjoint operators which are "near" to selfadjoint and to unitary ones, *Mathematical Notes*, **33**, 81-84.

[8] Gil', M. I. (1992). On an estimate for the norm of a function of a quasi-hermitian operator, *Studia Mathematica*, **103 (1)**, 17-24.

[9] Gil', M. I. (1993a). On inequalities for eigenvalues of matrices, *Linear Algebra and its Applications*, **184**, 201-206.

[10] Gil', M. I. (1993b). Estimates for norm of matrix-valued functions , *Linear and Multilinear Algebra*, **35**, 65-73.

[11] Gil', M. I. (1995). *Norm Estimations for Operator-valued Functions and Applications*. Marcel Dekker, New York.

[12] Henrici, P. (1962). Bounds for iterates, inverses, spectral variation and field of values of nonnormal matrices. *Numerische Mathematik*, **4**, 24-39.

[13] Kress, R., De Vries, H. L. and Wegmann, R. (1974). On non-normal matrices. *Linear Algebra Appl.*, **8**, 109-120.

[14] Marcus, M. and Minc, H. (1964). *A Survey of Matrix Theory and Matrix Inequalities*. Allyn and Bacon, Boston.

3. Invertibility of Finite Matrices

The present chapter deals with various types of invertibility conditions for finite matrices. In particular, we improve the classical Levy-Desplanques theorem and other well-known invertibility results for matrices that are close to triangular ones.

3.1 Preliminary Results

For a matrix
$$A = (a_{jk})_{j,k=1}^n \ \ (n > 1),$$
put
$$V_+ = \begin{pmatrix} 0 & a_{12} & \dots & a_{1n} \\ 0 & 0 & \dots & a_{2n} \\ . & \dots & . & . \\ 0 & 0 & \dots & 0 \end{pmatrix}, \ V_- = \begin{pmatrix} 0 & \dots & 0 & 0 \\ a_{21} & \dots & 0 & 0 \\ . & \dots & . & \\ a_{n1} & \dots & a_{n,n-1} & 0 \end{pmatrix}$$
and
$$D = diag\,(a_{11}, a_{22}, ..., a_{nn}).$$

So $A = D + V_+ + V_-$. *In this chapter it is assumed that all the diagonal entries are nonzero.* So
$$d_0 := \min_{k=1,...,n} |a_{jj}| > 0.$$

In the present section $\|.\|$ is an arbitrary norm in $\mathbf{C}^n$. Recall that I is the unit operator. Set
$$W_\pm = D^{-1}V_\pm.$$

Theorem 3.1.1 *With the notation*

$$J(W_\pm) := \|(I - W_\pm)^{-1}\|,$$

let

$$\nu(A) \equiv \max\{\frac{1}{J(W_+)} - \|W_-\|, \frac{1}{J(W_-)} - \|W_+\|\} > 0. \qquad (1.1)$$

Then A is invertible and the inverse matrix satisfies the inequality

$$\|A^{-1}\| \leq \frac{\|D^{-1}\|}{\nu(A)}. \qquad (1.2)$$

To prove this theorem we need the following two simple lemmas.

Lemma 3.1.2 *Let*

$$\psi_0 \equiv \|(I + W_-)^{-1}W_+\| < 1.$$

Then the operator $B \equiv I - W_- + W_+$ is boundedly invertible. Moreover,

$$\|B^{-1}\| \leq \frac{J(W_-)}{1 - \psi_0}.$$

Proof: Clearly, operators $W_\pm$ are nilpotent. So operators $I + W_\pm$ are invertible. We have

$$B = (I + W_-)(I + (I + W_-)^{-1}W_+).$$

Moreover,

$$\|(I + (I + W_-)^{-1}W_+)^{-1}\| \leq \sum_{k=0}^{\infty} \|((I + W_-)^{-1}W_+)^k\| \leq$$

$$\sum_{k=0}^{\infty} \psi_0^k = (1 - \psi_0)^{-1}.$$

Thus,

$$\|B^{-1}\| \leq \|(I + (I + W_-)^{-1}W_+)^{-1}\|\|(I + W_-)^{-1}\| \leq$$
$$(1 - \psi_0)^{-1}\|(I + W_-)^{-1}\|,$$

as claimed. □

Lemma 3.1.3 *Let at least one of the following inequalities:*

$$\|W_+\|J(W_-) < 1 \qquad (1.3)$$

or

$$\|W_-\|J(W_+) < 1 \qquad (1.4)$$

hold. Then relation (1.1) is valid. Moreover, the operator $B \equiv I + W_- + W_+$ is invertible and

$$\|B^{-1}\| \leq \frac{1}{\nu(A)}.$$

Proof: If condition (1.3) holds, then Lemma 3.1.2 yields the inequality

$$\|B^{-1}\| \leq \frac{J(W_-)}{1 - \|W_+\|J(W_-)} = \frac{1}{J^{-1}(W_-) - \|W_+\|}. \tag{1.5}$$

Interchanging W_- and W_+, under condition (1.4), we get

$$\|B^{-1}\| \leq \frac{1}{J^{-1}(W_+) - \|W_-\|}.$$

This relation and (1.5) yield the required result. $\square$

Proof of Theorem 3.1.1: Clearly, condition (1.1) implies that at least one of inequalities (1.3) or (1.4) holds. But

$$A = D + V_+ + V_- = D(I + W_+ + W_-) = DB.$$

Now the required result follows from Lemma 3.1.3.

3.2 l^p-Norms of Powers of Nilpotent Matrices

Recall that

$$\|h\|_p = [\sum_{k=1}^{n} |h_k|^p]^{1/p} \quad (h = (h_k) \in \mathbf{C}^n; \ 1 < p < \infty).$$

In the present and next sections $\|A\|_p$ is an operator norm of a matrix A with respect to the vector norm $\|.\|_p$. Put

$$\gamma_{n,m,p} = [C_{n-1}^m (n-1)^{-m}]^{1/p} \quad (m = 1, ..., n-1) \text{ and } \gamma_{n,0,p} = 1,$$

where $C_n^m = n!/(n-m)!m!$ are the binomial coefficients. Note that

$$\gamma_{n,m,p}^p = \frac{(n-1)!}{(n-1)^m(n-1-m)!m!} \leq \frac{(n-1)...(n-m)}{(n-1)^m m!}.$$

Hence,

$$\gamma_{n,m,p}^p \leq \frac{1}{m!} \quad (m = 1, ...n-1).$$

Lemma 3.2.1 *For any upper triangular nilpotent matrix*

$$V_+ = (a_{jk})_{j,k=1}^n \quad \text{with } a_{jk} = 0 \ (1 \leq k \leq j \leq n)$$

the inequality

$$\|V_+^m\|_p \leq \gamma_{n,m,p} M_p^m(V_+) \quad (m = 1, \ldots, n-1) \tag{2.1}$$

is valid with the notation

$$M_p(V_+) = (\sum_{j=1}^{n-1} [\sum_{k=j+1}^{n} |a_{jk}|^q]^{p/q})^{1/p} \quad (p^{-1} + q^{-1} = 1).$$

Proof: For a natural $s = 1, ..., n - 1$, denote

$$\|x\|_{p,s} = (\sum_{k=s}^{n} |x_k|^p)^{1/p},$$

where x_k are the coordinates of a vector $x \in \mathbf{C}^n$. We can write out

$$\|V_+x\|_{p,s}^p = \sum_{j=s}^{n-1} |\sum_{k=j+1}^{n} a_{jk}x_k|^p.$$

By Hölder's inequality,

$$\|V_+x\|_{p,s}^p \leq \sum_{j=s}^{n-1} h_j \|x\|_{p,j+1}^p, \tag{2.2}$$

where

$$h_j = [\sum_{k=j+1}^{n} |a_{jk}|^q]^{p/q}.$$

Similarly,

$$\|V_+^2x\|_{p,s}^p = \sum_{j=s}^{n-1} |\sum_{k=j+1}^{n} a_{jk}(V_+x)_k|^p \leq$$

$$\sum_{j=s}^{n-1} h_j \|V_+x\|_{p,j+1}^p.$$

Here $(V_+x)_k$ are the coordinates of V_+x. Taking into account (2.2), we obtain

$$\|V_+^2x\|_{p,s}^p \leq \sum_{j=s}^{n-1} h_j \sum_{k=j+1}^{n-1} h_k \|x\|_{p,k+1}^p =$$

$$\sum_{s \leq j < k \leq n-1} h_j h_k \|x\|_{p,k+1}^p.$$

Therefore,

$$\|V_+^2\|_p^p = \|V_+^2\|_{p,1}^p \leq \sum_{1 \leq j < k \leq n-1} h_j h_k.$$

Repeating these arguments, we arrive at the inequality

$$\|V_+^m\|_p^p \leq \sum_{1 \leq k_1 < k_2 < ... < k_m \leq n-1} h_{k_1} ... h_{k_m}. \tag{2.3}$$

Since,

$$\sum_{j=1}^{n} h_j = \sum_{j=1}^{n} [\sum_{k=j+1}^{n} |a_{jk}|^q]^{p/q} = M_p^p(V_+),$$

due to Lemma 2.4.1 and (2.3) we have

$$\|V_+^m\|_p^p \leq M_p^{mp}(V_+)C_{n-1}^m(n-1)^{-m},$$

as claimed. $\square$

Similarly we can prove

Lemma 3.2.2 *For any lower triangular nilpotent matrix*

$$V_- = (a_{jk})_{j,k=1}^n \text{ with } a_{jk} = 0 \ (1 \leq j \leq k \leq n).$$

the inequality

$$\|V_-^m\|_p \leq \gamma_{n,m,p}M_p^m(V_-) \ (m = 1, \ldots, n-1)$$

is valid, where

$$M_p(V_-) = (\sum_{j=2}^n [\sum_{k=1}^{j-1} |a_{jk}|^q]^{p/q})^{1/p}.$$

Consider the case $p = 2$. The Euclidean norm $\|.\|_2$ is invariant with respect to an orthogonal basis. Moreover,

$$M_2^2(V_+) = Trace \ V_+^* V_+ = \sum_{j=1}^{n-1} \sum_{k=j+1}^n |a_{jk}|^2 = N_2^2(V_+),$$

where $N_2(.) = N(.)$ is the Hilbert-Schmidt norm. Due to Lemma 3.2.1 we have

Corollary 3.2.3 *Any $n \times n$-nilpotent matrix V satisfies the inequalities*

$$\|V^m\|_2 \leq \gamma_{n,m,2}N_2^m(V) \ (m = 1, ..., n-1).$$

Thus, Lemma 3.2.1 gives us the new proof of Lemma 2.5.1, since $\gamma_{n,m,2} = \gamma_{n,m}$.

3.3   Invertibility in the Norm $\|.\|_p$ $(1 < p < \infty)$

Recall that A, d_0 and $V_\pm$ are defined in Section 3.1; $\gamma_{n,m,p}$ and $M_p(V_\pm)$ are defined in the previous section. In addition, $W_\pm = D^{-1}V_\pm$. So

$$M_p(W_-) = (\sum_{j=2}^n [\sum_{k=1}^{j-1} \frac{|a_{jk}|^q}{|a_{jj}|^q}]^{p/q})^{1/p}$$

and

$$M_p(W_+) = (\sum_{j=1}^{n-1} [\sum_{k=j+1}^n \frac{|a_{jk}|^q}{|a_{jj}|^q}]^{p/q})^{1/p} \ (p^{-1} + q^{-1} = 1).$$

Theorem 3.3.1 *With the notation*

$$J_p(W_\pm) := \sum_{k=0}^{n-1} \gamma_{n,k,p} M_p^k(W_\pm),$$

let

$$\nu_p(A) := \max\{\frac{1}{J_p(W_+)} - \|W_-\|_p, \frac{1}{J_p(W_-)} - \|W_+\|_p\} > 0.$$

Then A is invertible and the inverse matrix satisfies the inequality

$$\|A^{-1}\|_p \leq \frac{1}{d_0 \nu_p(A)}.$$

Proof: Clearly,

$$\|(I - W_\pm)^{-1}\|_p \leq \sum_{k=0}^{n-1} \|W_\pm^k\|_p.$$

Lemmas 3.2.1 and 3.2.2 imply

$$\|(I - W_\pm)^{-1}\|_p \leq J_p(W_\pm).$$

Now Theorem 3.1.1 yields the required result. □

3.4   Invertibility in the Norm $\|.\|_\infty$

For a matrix $A = (a_{jk})_{j=1}^n$ take the norm

$$\|A\|_\infty \equiv \max_{j=1,\ldots,n} \sum_{k=1}^n |a_{jk}|.$$

Recall that d_0 is defined in Section 3.1. Under the condition $d_0 > 0$, introduce the notation:

$$\tilde{v}_k := \max_{j=1,\ldots,k-1} |a_{jk}| \quad (k = 2,\ldots,n);$$

$$\tilde{w}_k := \max_{j=k+1,\ldots,n} |a_{jk}| \quad (k = 1,\ldots,n-1),$$

$$m_{up}(A) := \prod_{k=2}^n (1 + \frac{\tilde{v}_k}{|a_{kk}|}) \text{ and } m_{low}(A) := \prod_{k=1}^{n-1} (1 + \frac{\tilde{w}_k}{|a_{kk}|}).$$

Theorem 3.4.1 *Let the condition*

$$m_{up}(A) m_{low}(A) < m_{up}(A) + m_{low}(A) \tag{4.1}$$

be fulfilled. Then matrix A is invertible and the inverse matrix satisfies the inequality

$$\|A^{-1}\|_\infty \leq \frac{m_{up}(A) m_{low}(A)}{(m_{up}(A) + m_{low}(A) - m_{up}(A) m_{low}(A)) d_0}. \tag{4.2}$$

The proof of this theorem is divided into a series of lemmas which are presented in the next section. Note that condition (4.1) is equivalent to the following one:

$$\theta(A) := (m_{up}(A) - 1)(m_{low}(A) - 1) < 1. \tag{4.3}$$

Inequality (4.2) can be written as

$$\|A^{-1}\|_\infty \leq \frac{m_{up}(A)m_{low}(A)}{d_0(1 - \theta(A))}. \tag{4.4}$$

If matrix A is triangular and has nonzero diagonal entries, then (4.1) obviously holds.

3.5   Proof of Theorem 3.4.1

Recall that $V_\pm$ and D are introduced in Section 3.1, and $W_\pm = V_\pm D^{-1}$. In this section $\|A\|$ is the operator norm of A with respect to an arbitrary vector norm.

Lemma 3.5.1 *Let the condition*

$$\tilde{\theta}_0 \equiv \|\sum_{j,k=1}^{n-1} (-1)^{k+j} W_-^k W_+^j\| < 1 \tag{5.1}$$

hold. Then A is invertible and the inverse matrix satisfies the inequality

$$\|A^{-1}\| \leq \|D^{-1}\|\|(I + W_-)^{-1}\|\|(I + W_+)^{-1}\|(1 - \tilde{\theta}_0)^{-1}. \tag{5.2}$$

Proof: Clearly,

$$A = D + W_- + W = (I + W_- + W_+)D =$$

$$[(I + W_-)(I + W_+) - W_-W_+]D.$$

But W_+ and W_- are nilpotent:

$$W_-^n = W_+^n = 0. \tag{5.3}$$

So the operators, $I + W_-$ and $I + W_+$ are invertible:

$$(I + W_-)^{-1} = \sum_{k=0}^{n-1}(-1)^k W_-^k,$$

$$(I + W_+)^{-1} = \sum_{k=0}^{n-1}(-1)^k W_+^k. \tag{5.4}$$

Thus

$$A = (I + W_-)[I - (I + W_-)^{-1}W_-W_+(I + W_+)^{-1}](I + W_+)D.$$

Thanks to (5.4) we have

$$(I + W_-)^{-1}W_- = \sum_{k=1}^{n-1}(-1)^{k-1}W_-^k, \ W_+(I + W_+)^{-1} =$$

$$\sum_{k=1}^{n-1}(-1)^{k-1}W_+^k.$$

So

$$A = (I + W_-)[I - \sum_{j,k=1}^{n-1}(-1)^{k+j}W_-^kW_+^j](I + W_+)D.$$

Therefore, if (5.1) holds then A is invertible. Moreover

$$A^{-1} = D^{-1}(I + W_+)^{-1}[I - \sum_{j,k=1}^{n-1}(-1)^{k+j}W_-^kW_+^j]^{-1}(I + W_-)^{-1}. \quad (5.5)$$

Condition (5.1) yields

$$\|[I - \sum_{j,k=1}^{n-1}(-1)^{k+j}W_-^kW_+^j]^{-1}\| \le (1 - \tilde{\theta}_0)^{-1}.$$

Now inequality (5.2) is due to (5.5). $\square$

Denote

$$\tilde{m}(V_+) = \prod_{k=2}^{n}(1 + \tilde{v}_k) \text{ and } \tilde{m}(V_-) = \prod_{k=1}^{n-1}(1 + \tilde{w}_k).$$

Lemma 3.5.2 *The inequalities*

$$\|(I - V_+)^{-1}\|_\infty \le \tilde{m}(V_+) \quad (5.6)$$

and

$$\|(I - V_-)^{-1}\|_\infty \le \tilde{m}(V_-) \quad (5.7)$$

are valid.

Proof: Let Q_k be the projectors onto the standard basis:

$$Q_k h = (h_1, h_2, ..., h_k, 0, 0, ..., 0) \ (k = 1, ..., n), \ Q_0 = 0$$

for an arbitrary vector $h = (h_1, ..., h_n) \in \mathbf{C}^n$. Clearly, Q_k project onto the invariant subspaces of V_+. So according to Lemma 2.10.1,

$$(I - V_+)^{-1} = \overrightarrow{\prod_{2 \le k \le n}} (I + V_+\Delta Q_k), \text{ where } \Delta Q_k = Q_k - Q_{k-1}. \quad (5.8)$$

It is not hard to check that

$$\|V_+ \Delta Q_k\|_\infty = \tilde{v}_k.$$

Now inequality (5.6) follows from (5.8). Further, define a projector $\tilde{Q}_k$ by

$$\tilde{Q}_k h = (0, 0, ..., h_{n-k+1}, h_{n-k+2}, ..., h_n)$$

$$(k = 1, ..., n), \ \tilde{Q}_0 = 0.$$

Simple calculation show that $\tilde{Q}_k$ project onto invariant subspaces of V_-. So according Lemma 2.10.1,

$$(I - V_-)^{-1} = \overset{\rightarrow}{\prod_{2 \le k \le n}} (I + V_- \Delta \tilde{Q}_k),$$

$$(\Delta \tilde{Q}_k = \tilde{Q}_k - \tilde{Q}_{k-1}). \tag{5.9}$$

It is not hard to check that $\|V_- \Delta \tilde{Q}_k\|_\infty = \tilde{w}_{n-k+1}$. Now inequality (5.7) follows from (5.9). $\square$

Lemma 3.5.3 *The inequalities*

$$\|\sum_{k=1}^{n-1}(-1)^k V_+^k\|_\infty \le \tilde{m}(V_+) - 1 \ and \ \|\sum_{k=1}^{n-1}(-1)^k V_-^k\|_\infty \le \tilde{m}(V_-) - 1 \tag{5.10}$$

are valid.

Proof: Let $B = (b_{jk})_{k=1}^n$ be a nonnegative matrix with the property

$$Bh \ge h \tag{5.11}$$

for any nonnegative $h \in \mathbf{C}^n$. Then $b_{jj} \ge 1$ $(j = 1, ..., n)$. Hence,

$$\|B - I\|_\infty = \max_{j=1,...,n} [\sum_{k=1}^n b_{jk} - \delta_{jk}] = \|B\|_\infty - 1. \tag{5.12}$$

Here δ_{jk} is the Kronecker symbol. Furthermore, since V_+ is nilpotent,

$$\|\sum_{k=1}^{n-1}(-1)^k V_+^k\|_\infty \le \|\sum_{k=1}^{n-1}|V_+|^k\|_\infty =$$

$$\|(I - |V_+|)^{-1} - I\|_\infty$$

where $|V_+|$ is the matrix whose entries are the absolute values of the entries of V. Moreover, clearly,

$$\sum_{k=0}^{n-1}|V_+|^k h \ge h$$

for any nonnegative $h \in \mathbf{C}^n$. So according to (5.11) and (5.12),

$$\| \sum_{k=1}^{n-1} (-1)^k V_+^k \|_\infty \leq \| \sum_{k=1}^{n-1} |V_+|^k \|_\infty =$$

$$\| (I - |V_+|)^{-1} - I \|_\infty$$

Since

$$\tilde{m}(V_+) = \tilde{m}(|V_+|),$$

equality (5.6) with $V_+ = |V_+|$ yields the first inequality (5.10). Similarly, the second inequality (5.10) can be proved. $\square$

Proof of Theorem 3.4.1: Due to Lemma 3.5.2,

$$\| (I + W_-)^{-1} \|_\infty \leq m_{up}(A), \; \| (I + W_+)^{-1} \|_\infty \leq m_{low}(A).$$

Lemma 3.5.3 yields

$$\| \sum_{k=1}^{n-1} (-1)^k W_-^k \|_\infty \leq m_{up}(A) - 1,$$

and

$$\| \sum_{k=1}^{n-1} (-1)^k W_+^k \|_\infty \leq m_{low}(A) - 1.$$

Hence,

$$\| \sum_{j,k=1}^{n-1} (-1)^{k+j} W_-^k W_+^j \|_\infty \leq \theta(A).$$

Now the required result follows directly from Lemma 3.5.1. $\square$

3.6   Positive Invertibility of Matrices

For $h = (h_k), w = (w_k) \in \mathbf{R}^n$, we write $h \geq (>) g$, if $h_k \geq (>) w_k$, $(k = 1, ..., n)$. A matrix A is (non-negative) positive: $A > (\geq) 0$, if all its entries are (non-negative) positive. For matrices A, B we write $A > (\geq) B$ if $A - B > (\geq) 0$.

Again V_+, V_- and D are the upper triangular, lower triangular, and diagonal parts of matrix $A = (a_{jk})_{j,k=1}^n$, respectively (see Section 3.1).

Matrix $A = (a_{jk})_{j=1,k=1}^n$ is called a Z-matrix, if the conditions

$$a_{jk} \leq 0 \; \text{ for } j \neq k, \; \text{ and } a_{kk} > 0 \; (j,k = 1, ..., n) \qquad (6.1)$$

hold. That is, $D > 0, V_\pm \leq 0$.

Lemma 3.6.1 *Let A be a Z-matrix, and let condition (4.1) hold. Then A is positively invertible and the inverse operator satisfies the inequalities (4.2) and*

$$A^{-1} \geq D^{-1}. \tag{6.2}$$

Proof: As it was proved in the previous section, condition (4.1) implies the invertibility of A. Moreover, relation (5.5) holds. But $W_{\pm} \leq 0$. So

$$(-1)^{k+j} W_-^k W_+^j \geq 0$$

and thus

$$[I - \sum_{j,k=1}^{n-1} (-1)^{k+j} W_-^k W_+^j]^{-1} \geq 0,$$

since

$$\| \sum_{j,k=1}^{n-1} (-1)^{k+j} W_-^k W_+^j \| \leq \theta_0 < 1.$$

In addition,

$$(I + W_{\pm})^{-1} = \sum_{k=1}^{n-1} (-1)^k W_{\pm}^k \geq 0.$$

Now (5.5) implies (6.2). Inequality (4.2) is due to Theorem 3.4.1. $\square$

3.7   Positive Matrix-Valued Functions

We will call A an *anti-Hurwitz matrix* if its spectrum $\sigma(A)$ lies in the open right half-plane:

$$\beta(A) := \inf\ Re\ \sigma(A) > 0.$$

For an anti-Hurwitzian matrix A, define the matrix-valued function $F(A)$ by

$$F(A) = \int_0^\infty e^{-At} f(t) dt, \tag{7.1}$$

where

$$e^{-\beta(A)t} f(t) \in L^1[0, \infty). \tag{7.2}$$

Lemma 3.7.1 *Let A be an anti-Hurwitzian Z-matrix. In addition, let relations (7.1), (7.2) hold, and $f(t) \geq 0$ for almost all $t \geq 0$. Then*

$$F(A) \geq F(D) \geq 0. \tag{7.3}$$

Proof: Since $D > 0$ and $V_\pm \leq 0$, we have

$$e^{-At} = \lim_{n \to \infty} (I - An^{-1})^{nt} = \lim_{n \to \infty} (I - n^{-1}D - n^{-1}V)^{nt} \geq$$

$$\lim_{n \to \infty} (I - n^{-1}D)^{nt} = e^{-Dt},$$

where I is the unit matrix. Thus

$$F(A) = \int_0^\infty e^{-At} f(t) dt \geq \int_0^\infty e^{-Dt} f(t) dt = F(D).$$

As claimed. □

In, particular, let A be an anti-Hurwitzian matrix, and

$$f(t) \equiv \sum_{k=1}^\infty \frac{a_k t^{s_k}}{\Gamma(s_k + 1)}$$

$$(s_k = const > 0, \ a_k = const \geq 0, k = 1, 2, ...),$$

where $\Gamma(.)$ is the Euler Gamma function. Then

$$F(A) = \sum_{k=1}^\infty a_k A^{-s_k - 1}, \tag{7.4}$$

provided the series converges. The following functions are examples of the functions defined by (7.4):

$$A^{-\nu}, \ A^{-\nu} e^{b_0 A^{-1}} \ (b_0 = const > 0; \ 0 < \nu \leq 1).$$

Lemma 3.7.2 *Let A be a Z-matrix. Then it is anti-Hurwitzian and satisfies the inequality (6.2) if and only if it is positively invertible.*

Proof: Let A be anti-Hurwitzian. Then due to Lemma 3.7.1,

$$A^{-1} = \int_0^\infty e^{-At} dt \geq D^{-1} > 0.$$

Conversely, let A be positively invertible. Put $T = -V_+ - V_-$. Then for any λ with $Re \ \lambda \geq 0$,

$$|(A + \lambda)h| = |(D + \lambda - T)h| \geq |(D + \lambda)h| - T|h| \geq D|h| - T|h| = A|h|.$$

Here $|h|$ means the vectors whose coordinates are the absolute values of the coordinates of h. This proves the result. □
Lemma 3.6.1 and the previous lemma imply

Corollary 3.7.3 *Let F be defined by (7.1) with a non-negative function f satisfying (7.2). In addition, let A be a Z-matrix and condition (4.1) hold. Then relation (7.3) is valid.*

3.8 Notes

As it was mentioned, although excellent computer softwares are now available for eigenvalue computation, new results on invertibility and spectrum inclusion regions for finite matrices are still important, since computers are not very useful, in particular, for analysis of matrices dependent on parameters. So the problem of finding invertibility conditions and spectrum inclusion regions for finite matrices continues to attract attention of many specialists, cf. (Brualdi, 1982), (Farid, 1995 and 1998), (Gudkov, 1967), (Li and Tsatsomeros, 1997), (Tam et al., 1997) and references given therein.

Let $A = (a_{jk})$ be a complex $n \times n$-matrix $(n \geq 2)$ with the nonzero diagonal: $a_{kk} \neq 0$ $(k = 1, ..., n)$. Put

$$P_j = \sum_{k=1,\ k\neq j}^{n} |a_{jk}|.$$

The well-known Levy-Desplanques theorem states that if $|a_{jj}| > P_j$ $(j = 1, ..., n)$, then A is nonsingular. This theorem has been improved in many ways. For example, each of the following is known to be a sufficient condition for non-singularity of A:

(i) $|a_{ii}||a_{jj}| > P_j P_i$ $(i, j = 1, ..., n)$ (Marcus and Minc, 1964, p. 149).

(ii) $|a_{jj}| \geq P_j$ $(j = 1, ..., n)$, provided that at least one inequality is strict and A is irreducible (Marcus and Minc, 1964, p. 147).

(iii) $|a_{jj}| \geq r_j m_j$ $(j = 1, ..., n)$, where r_j are positive numbers satisfying

$$\sum_{k=1}^{n}(1 + r_k)^{-1} \leq 1 \text{ and } m_j = \max_{k\neq j} |a_{jk}| \text{ (see (Bailey and Crabtree, 1969).}$$

(iv) $|a_{jj}| > P_j^\epsilon Q_j^{1-\epsilon}$ $(j = 1, ..., n)$ where

$$0 \leq \epsilon \leq 1, \ Q_j = \sum_{k=1,\ k\neq j}^{n} |a_{kj}| \text{ (Marcus and Minc, 1964, p. 150) .}$$

Theorems 3.1.1, 3.3.1 and 3.4.1 yield new invertibility conditions which improve the mentioned results, when the considered matrices are close to triangular ones. Moreover, they give us estimates for different norms of the inverse matrices. Note that Theorems 3.1.1, 2.1.1 and 2.14.1 allow us to derive additional invertibility conditions in the terms of the Euclidean norm.

The material in Chapter 3 is based on the papers (Gil', 1997), (Gil', 1998) and (Gil', 2001).

References

[1] Bailey D. W. and D. E. Crabtree, (1969), Bounds for determinants, *Linear Algebra and Its Applications*, **2**, 303-309.

[2] Brualdi, R.A. (1982), Matrices, eigenvalues and directed graphs, *Linear and Multilinear Algebra*, **11**, 143-165

[3] Collatz, L. (1966). *Functional Analysis and Numerical Mathematics*. Academic press, New York and London.

[4] Farid, F.O. (1995), Criteria for invertibility of diagonally dominant matrices, *Linear Algebra and Its Applications*, **215**, 63-93.

[5] Farid, F.O. (1998), Topics on a generalization of Gershgorin's theorem, *Linear Algebra and Its Applications*, **268**, 91-116.

[6] Gil', M.I. (1997), A nonsingularity criterion for matrices, *Linear Algebra and Its Applications*, **253**, 79-87.

[7] Gil', M.I. (1998), On positive invertibility of matrices, *Positivity*, **2**, 165-170.

[8] Gil', M.I. (2001), Invertibility and positive invertibility of matrices, *Linear Algebra and and Its Applications*, **327**, 95-104

[9] Gudkov, V. V. (1967). On a certain test for nonsingularity of matrices, *Latvian Math. Yearbook, 1965*, 385-390, (*Math. Reviews, 33), review 1323*)

[10] Horn, R. A. and Johnson Ch. R. (1991). *Topics in Matrix Analysis*, Cambridge, Cambridge University Press.

[11] Krasnosel'skii, M. A., Lifshits, J. and A. Sobolev. (1989). *Positive Linear Systems. The Method of Positive Operators*, Heldermann Verlag, Berlin.

[12] Li B. and Tsatsomeros, M.J. (1997), Doubly diagonally dominant matrices, *Linear Algebra and Its Applications*, **261**, 221-235.

[13] Marcus M. and Minc, H. (1964). *A Survey of Matrix Theory and Matrix Inequalities*, Allyn and Bacon, Boston.

[14] Tam, B.S., Yang, S. and Zhang. X. (1997) Invertibility of irreducible matrices, *Linear Algebra and Its Applications*, **259**, 39-70

4. Localization of Eigenvalues of Finite Matrices

The present chapter is concerned with perturbations of finite matrices and bounds for their eigenvalues. In particular, we improve the classical Gershgorin result for matrices, which are "close" to triangular ones. In addition, we derive upper and lower estimates for the spectral radius. Under some restrictions, these estimates improve the Frobenius inequalities. Moreover, we present new conditions for the stability of matrices, which supplement the Rohrbach theorem.

4.1 Definitions and Preliminaries

In this chapter, $\|.\|$ is an arbitrary norm in $\mathbf{C}^n$, A and B are $n \times n$-matrices having eigenvalues $\lambda_1(A), ..., \lambda_n(A)$ and $\lambda_1(B), ..., \lambda_n(B)$, respectively, and $q = \|A - B\|$.

We recall some well-known definitions from matrix perturbation theory (see (Stewart and Sun, 1990)). *The spectral variation of B with respect to A is*

$$sv_A(B) := \max_i \min_j |\lambda_i(B) - \lambda_j(A)|.$$

The Hausdorff distance between the spectra of A and B is

$$hd(A, B) := \max\{sv_A(B), sv_B(A)\}.$$

The matching distance between eigenvalues of A and B is

$$md(A, B) := \min_\pi \max_i |\lambda_{\pi(i)}(B) - \lambda_i(A)|,$$

where π is taken over all permutations of $\{1, 2, ..., n\}$.

Recall also that $\sigma(A)$ denotes the spectrum of A.

Lemma 4.1.1 *For any $\mu \in \sigma(B)$, we have either $\mu \in \sigma(A)$, or*

$$q\|R_\mu(A)\| \geq 1. \tag{1.1}$$

Proof: Suppose that the inequality

$$q\|R_\mu(A)\| < 1. \tag{1.2}$$

holds. We can write $R_\mu(A) - R_\mu(B) = R_\mu(B)(B - A)R_\mu(A)$. This yields

$$\|R_\mu(A) - R_\mu(B)\| \leq \|R_\mu(B)\| q \|R_\mu(A)\|. \tag{1.3}$$

Thus, (1.2) implies

$$\|R_\mu(B)\| \leq \|R_\mu(A)\|(1 - q\|R_\mu(A)\|)^{-1}.$$

That is, μ is a regular point of B. This contradiction proves the result. $\square$

Lemma 4.1.2 *Assume that*

$$\|R_\lambda(A)\| \leq \phi(\rho^{-1}(A, \lambda)) \text{ for all regular } \lambda \text{ of } A, \tag{1.4}$$

where $\phi(x)$ is a monotonically increasing non-negative function of a non-negative variable x, such that $\phi(0) = 0$ and $\phi(\infty) = \infty$. Then the inequality

$$sv_A(B) \leq z(\phi, q) \tag{1.5}$$

is true, where $z(\phi, q)$ is the extreme right-hand (positive) root of the equation

$$1 = q\phi(1/z). \tag{1.6}$$

Proof: Due to Lemma 4.1.1 and condition (1.4),

$$1 \leq q\phi(\rho^{-1}(A, \lambda)) \text{ for all } \lambda \in \sigma(B).$$

Since $\phi(x)$ monotonically increases, $z(\phi, q)$ is a unique positive root of (1.6) and $\rho(A, \lambda) \leq z(\phi, q)$. Thus, the inequality (1.5) is valid. $\square$

4.2 Perturbations of Multiplicities and Matching Distance

Put

$$\Omega(c, r) \equiv \{z \in \mathbf{C} : |z - c| \leq r\} \quad (c \in \mathbf{C}, r > 0).$$

Lemma 4.2.1 *Under condition (1.4), let A have an eigenvalue $\lambda(A)$ of the algebraic multiplicity ν and the distance from $\lambda(A)$ to the rest of $\sigma(A)$ is equal to $2d$, i.e.*

$$distance\{\lambda(A), \sigma(A)/\lambda(A)\} = 2d. \tag{2.1}$$

In addition, for a positive number $a \leq d$, let

$$q\phi(1/a) < 1. \tag{2.2}$$

Then in $\Omega(\lambda(A), a)$ operator B has eigenvalues whose total algebraic multiplicity is equal to ν.

Proof: This result is a particular case of the well-known Theorem 3.18 (Kato, 1966, p. 215). $\square$

Since ϕ is a nondecreasing function, comparing (2.2) with (1.6), we get

Corollary 4.2.2 *Let A have an eigenvalue $\lambda(A)$ of the algebraic multiplicity ν. In addition, under conditions (1.4) and (2.1), let the extreme right-hand root $z(\phi, q)$ of equation (1.6) satisfies the inequality*

$$z(\phi, q) \leq d. \tag{2.3}$$

Then in $\Omega(\lambda(A), z(\phi, q))$ operator B has eigenvalues whose total algebraic multiplicity is equal to ν.

Let $\theta_1, ..., \theta_{n_1}$ $(n_1 \leq n)$ be different eigenvalues of A, i.e.,

$$\tilde{d}(A) := \min\{|\theta_j - \theta_k| : j \neq k; \; j, k = 1, .., n_1\} > 0 \tag{2.4}$$

and

$$\sum_{k=1}^{n_1} \nu_k = n,$$

where ν_k is the multiplicity of the eigenvalue θ_k.

By virtue of Lemma 4.2.1, we arrive at the following result.

Lemma 4.2.3 *Let condition (1.4) hold. In addition, for a positive number $a \leq \tilde{d}(A)$, let condition (2.2) is valid. Then all the eigenvalues of operator B lie in the set*

$$\cup_{k=1}^{n_1} \Omega(\theta_k, a)$$

and thus, $md\,(A, B) \leq a$.

Moreover, Corollary 4.2.2 implies

Corollary 4.2.4 *Under conditions (1.4) and (2.4), let the extreme right-hand root $z(\phi, q)$ of equation (1.6) satisfies the inequality $z(\phi, q) \leq \tilde{d}(A)$. Then all the eigenvalues of operator B lie in the set*

$$\cup_{k=1}^{n_1} \Omega(\theta_k, z(\phi, q))$$

and thus $md\,(A, B) \leq z(\phi, q)$.

4.3   Perturbations of Eigenvectors and Eigenprojectors

Under condition (1.4), let A have an eigenvalue $\lambda(A)$ of the algebraic multiplicity ν and (2.1) hold. Let $\partial\Omega$ be the boundary of $\Omega(\lambda(A), d)$:

$$\partial\Omega \equiv \{z \in \mathbf{C} : |z - \lambda(A)| = d\}.$$

Put

$$P(A) = -\frac{1}{2\pi i} \int_{\partial\Omega} R_\lambda(A)d\lambda \qquad (3.1)$$

and

$$P(B) = -\frac{1}{2\pi i} \int_{\partial\Omega} R_\lambda(B)d\lambda. \qquad (3.2)$$

That is, both $P(A)$ and $P(B)$ are the projectors onto the eigenspaces of A and B, respectively, corresponding to points of the spectra, which lie in $\Omega(\lambda(A), d)$.

Lemma 4.3.1 *Let A satisfy condition (1.4) and have an eigenvalue $\lambda(A)$ of the algebraic multiplicity ν, such that the conditions (2.1) and*

$$q\phi(1/d) < 1 \qquad (3.3)$$

hold. Then $dim\ P(A) = dim\ P(B)$. Moreover,

$$\|P(A) - P(B)\| \leq \frac{qd\phi^2(1/d)}{1 - q\phi(1/d)}. \qquad (3.4)$$

Proof: Thanks to Lemma 4.2.1 $dim\ P(A) = dim\ P(B)$. From (1.3) and (1.4) it follows that

$$\|R_\lambda(B)\| \leq \|R_\lambda(A)\|(1 - q\phi(1/d))^{-1} \leq \phi(1/d)(1 - q\phi(1/d))^{-1} \quad (\lambda \in \partial\Omega)$$

and

$$\|P(A) - P(B)\| \leq \frac{1}{2\pi} \int_{\partial\Omega} \|R_\lambda(A) - R_\lambda(B)\| |d\lambda| \leq$$

$$\frac{1}{2\pi} \int_{\partial\Omega} \|R_\lambda(B)\| q\phi(1/d)|d\lambda| \leq \frac{q\phi^2(1/d)d}{1 - q\phi(1/d)},$$

as claimed. $\square$

We will say that an eigenvalue of a linear operator is *a simple eigenvalue*, if its algebraic multiplicity is equal to one. The eigenvector corresponding to a simple eigenvalue will be called a *simple eigenvector*.

Lemma 4.3.2 *Suppose A has a simple eigenvalue $\lambda(A)$, such that the relations (1.4), (2.1) and*

$$q[d\phi^2(1/d) + \phi(1/d)] < 1 \qquad (3.5)$$

hold. Then for the eigenvector e of A corresponding to $\lambda(A)$ with $\|e\| = 1$, there exists a simple eigenvector f of B with $\|f\| = 1$, such that

$$\|e - f\| \le 2\delta(1 - \delta)^{-1},$$

where

$$\delta \equiv \frac{qd\phi^2(1/d)}{1 - q\phi(1/d)}.$$

Proof: Firstly, note that condition (3.5) implies the relations (3.3) and $\delta < 1$, and B has in $\Omega(\lambda(A), d)$ a simple eigenvalue $\lambda(B)$ due to Lemma 4.2.1. Let $P(B)$ and $P(A)$ be defined by (3.1) and (3.2). Due to Lemma 4.3.1,

$$\|P(A) - P(B)\| \le \delta < 1.$$

Consequently, $P(B)e \ne 0$, since $P(A)e = e$. Thanks to the relation

$$BP(B)e = \lambda(B)P(B)e,$$

$P(B)e$ is an eigenvector of B. Let $N = \|P(B)e\|$. Then $f \equiv N^{-1}P(B)e$ is a normed eigenvector of B. So

$$e - f = P(A)e - N^{-1}P(B)e = e - N^{-1}e + N^{-1}(P(A) - P(B))e.$$

But

$$N \ge \|P(A)e\| - \|(P(A) - P(B))e\| \ge 1 - \delta.$$

Hence $N^{-1} \le (1 - \delta)^{-1}$ and

$$\|e - f\| \le (N^{-1} - 1)\|e\| + N^{-1}\|P(A) - P(B)\| \le$$
$$(1 - \delta)^{-1} - 1 + (1 - \delta)^{-1}\delta = 2\delta(1 - \delta)^{-1},$$

as claimed. $\square$

4.4   Perturbations of Matrices in the Euclidean Norm

4.4.1   Perturbations of eigenvalues

We recall that the quantities $\|.\|_2, g(A)$ and $\gamma_{n,k}$ are defined in Sections 1.2 and 2.1. In addition, put

$$q_2 := \|A - B\|_2.$$

The norm for matrices is understood in the sense of the operator norm.

Theorem 4.4.1 *Let A and B be $n \times n$-matrices. Then*

$$sv_A(B) \leq z(q_2, A), \tag{4.1}$$

where $z(q_2, A)$ is the extreme right-hand (unique nonnegative) root of the algebraic equation

$$z^n = q_2 \sum_{j=0}^{n-1} \gamma_{n,j} z^{n-j-1} g^j(A). \tag{4.2}$$

Proof: Theorem 2.1.1 gives us the inequality

$$\|R_\lambda(A)\|_2 \leq \sum_{k=0}^{n-1} \frac{g^k(A)\gamma_{n,k}}{\rho^{k+1}(A,\lambda)} \quad (\lambda \notin \sigma(A)). \tag{4.3}$$

Rewrite (4.2) as

$$1 = \sum_{j=0}^{n-1} \frac{\gamma_{n,j} g^j(A)}{z^{j+1}}.$$

Now the required result is due to Lemma 4.1.2 and (4.3). □
    Put

$$w_n = \sum_{j=0}^{n-1} \gamma_{n,j}.$$

Setting $z = g(A)y$ in (4.2) and applying Lemma 1.6.1, we have the estimate $z(q_2, A) \leq q_2 w_n$, provided

$$q_2 w_n \geq g(A) \tag{4.4}$$

and

$$z(q_2, A) \leq g^{1-1/n}(A)[q_2 w_n]^{1/n},$$

provided

$$q_2 w_n \leq g(A). \tag{4.5}$$

Now Theorem 4.4.1 ensures the following result.

Corollary 4.4.2 *Let condition (4.4) hold. Then $sv_A(B) \leq qw_n$. If condition (4.5) holds, then*

$$sv_A(B) \leq g^{1-1/n}(A)[qw_n]^{1/n}.$$

4.4.2   Perturbations of multiplicities and matching distance

For a positive scalar variable x, set

$$G_n(x) \equiv \sum_{j=0}^{n-1} \gamma_{n,j} x^{j+1} g^j(A).$$

Lemma 4.4.3 *Let A have an eigenvalue $\lambda(A)$ of the algebraic multiplicity ν and (2.1) hold. In addition, for a positive number $a < d$, let*

$$q_2 G_n(1/a) < 1. \tag{4.6}$$

Then in $\Omega(\lambda(A), a)$ operator B has eigenvalues whose total algebraic multiplicity is equal to ν.

Proof: This result follows from (4.3) and Lemma 4.2.1. □

Inequality (4.3) and Corollary 4.2.2 imply

Corollary 4.4.4 *Let A have an eigenvalue $\lambda(A)$ of the algebraic multiplicity ν. In addition, under condition (2.1), let the extreme right-hand root $z(q_2, A)$ of equation (4.2) satisfies the inequality $z(q_2, A) \leq d$. Then in $\Omega(\lambda(A), z(q_2, A))$ operator B has eigenvalues whose total algebraic multiplicity is equal to ν.*

Let $\theta_1, ..., \theta_{n_1}$ $(n_1 \leq n)$ be the different eigenvalues of A, again. By virtue of (4.3) and Lemma 4.2.3 we get the following result.

Lemma 4.4.5 *Under (2.4), for a positive number $a < \tilde{d}(A)$, let condition (4.6) is valid. Then all the eigenvalues of operator B lie in the set*

$$\cup_{k=1}^{n_1} \Omega(\theta_k, a)$$

and thus $md\,(A, B) \leq a$.

In addition, due to (4.3) and Corollary 4.2.4, we get

Corollary 4.4.6 *Under (2.4), let the extreme right-hand root $z(q_2, A)$ of equation (4.2) satisfies the inequality $z(q_2, A) \leq \tilde{d}(A)$. Then all the eigenvalues of operator B lie in the set*

$$\cup_{k=1}^{n_1} \Omega(\theta_k, z(q_2, A))$$

and thus $md\,(A, B) \leq z(q_2, A)$.

To estimate $z(q_2, A)$ we can apply Lemma 1.6.1.

4.4.3 Perturbations of eigenvectors and eigenprojectors in the Euclidean norm

Let A have an eigenvalue $\lambda(A)$ of the algebraic multiplicity ν and (2.1) hold. Define $P(A)$ and $P(B)$ as in Section 3.1. Recall that G_n and q_2 are defined in the previous two subsections.

Lemma 4.4.7 *Let A and B be linear operators in $\mathbf{C}^n$. In addition, let A have an eigenvalue $\lambda(A)$ of the algebraic multiplicity ν, such that the conditions (2.1) and*

$$q_2 G_n(1/d) < 1$$

hold. Then $\dim P(A) = \dim P(B)$. Moreover,

$$\|P(A) - P(B)\|_2 \leq \frac{q_2 G_n(1/d)d}{1 - q_2 G_n(1/d)}.$$

This result is due to (4.3) and Lemma 4.3.1.

Lemma 4.4.8 *Let A and B be linear operators in $\mathbf{C}^n$. Suppose A has a simple eigenvalue $\lambda(A)$, such that the relations (2.1) and*

$$q_2 [G_n^2(1/d)d + G_n(1/d)] < 1$$

hold. Then for the eigenvector e of A corresponding to $\lambda(A)$ with $\|e\|_2 = 1$, there exists the simple eigenvector f of B with $\|f\|_2 = 1$, such that

$$\|e - f\|_2 \leq 2\delta_2(1 - \delta_2)^{-1},$$

where

$$\delta_2 \equiv \frac{q_2 d G_n^2(1/d)}{1 - q_2 G_n(1/d)}.$$

This result is due to (4.3) and Lemma 4.3.2.

4.5 Upper Bounds for Eigenvalues in Terms of the Euclidean Norm

Let A be an $n \times n$-matrix with entries a_{jk} $(j, k = 1, ..., n)$. Recall that D, V_+ and V_- are the diagonal, upper nilpotent part and lower nilpotent part of A, respectively, (see Section 3.1), and

$$w_n = \sum_{j=0}^{n-1} \gamma_{n,j},$$

where $\gamma_{n,j}$ are defined in Section 2.1. Assume that $V_+ \neq 0, V_- \neq 0$ and denote

$$\mu_2(A) = w_n \min \{N^{-1}(V_+)\|V_-\|_2, \ N^{-1}(V_-)\|V_+\|_2\},$$

and

$$\delta_2(A) = w_n^{1/n} \min \{N^{1-1/n}(V_+)\|V_-\|_2^{1/n}, N^{1-1/n}(V_-)\|V_+\|_2^{1/n}\} \text{ if } \mu_2(A) \leq 1$$

$$\text{and } \delta_2(A) = \min \{\|V_-\|_2, \|V_+\|_2\} \text{ if } \mu_2(A) > 1.$$

Theorem 4.5.1 *All the eigenvalues of matrix $A = (a_{jk})_{j,k=1}^n$ lie in the union of the discs*

$$\{\lambda \in \mathbf{C} : |\lambda - a_{kk}| \le \delta_2(A)\}, \ k = 1, ..., n. \tag{5.1}$$

Proof: Take $A_+ = D + V_+$. Since A_+ is triangular,

$$\sigma(A_+) = \sigma(D) = \{a_{kk}, \ k = 1, ..., n\}. \tag{5.2}$$

Since $A - A_+ = V_-$, Corollary 4.4.2 implies

$$sv_{A_+}(A) \le N^{1-1/n}(V_+)[\|V_-\|_2 w_n]^{1/n}, \tag{5.3}$$

provided that $w_n\|V_-\|_2 \le N(V_+)$. Replace A_+ by $A_- = D + V_-$. Repeating the above procedure, we get

$$sv_{A_-}(A) \le N^{1-1/n}(V_-)[\|V_+\|_2 w_n]^{1/n},$$

provided that $w_n\|V_+\|_2 \le N(V_-)$. In addition, $\sigma(A_-) = \sigma(D)$. These relations with (5.2) and (5.3) complete the proof in the case $\mu_2(A) \le 1$. Te case $\mu_2(A) > 1$ is similarly considered $\square$

4.6   Lower Bounds for the Spectral Radius

Let $A = (a_{jk})$ be an $n \times n$-matrix. Recall that $r_s(A) = \sup |\sigma(A)|$ is the (upper) spectral radius;

$$\alpha(A) = \sup \ Re \ \sigma(A) \text{ and } r_l(A) = \inf |\sigma(A)|$$

is the inner (lower) spectral radius. Let V_+, V_- be the upper and lower triangular parts of A (see Section 3.1). Denote by $z(\nu)$ the unique positive root of the equation

$$z^n(A) = \nu(A) \sum_{k=0}^{n-1} g^k(A)\gamma_{n,k} z^{n-k-1} \tag{6.1}$$

where

$$\nu(A) = \min\{\|V_-\|_2, \ \|V_+\|_2\}.$$

Theorem 4.6.1 *Let $A = (a_{jk})_{j,k=1}^n$ be an $n \times n$-matrix. Then for any $k = 1, ..., n$, there is an eigenvalue μ_0 of A, such that*

$$|\mu_0 - a_{kk}| \le z(\nu). \tag{6.2}$$

Moreover, the following inequalities are true:

$$r_s(A) \ge \max\{0, \max_{k=1,...,n} |a_{kk}| - z(\nu)\}, \tag{6.3}$$

$$r_l(A) \le \min_{k=1,...,n} |a_{kk}| + z(\nu), \tag{6.4}$$

and

$$\alpha(A) \ge \max_{k=1,...,n} Re \ a_{kk} - z(\nu). \tag{6.5}$$

Proof: Take $A_+ = D + V_+$. So relation (5.2) holds and $A - A_+ = V_-$. Theorem 4.3.1 gives us the inequality

$$sv_A(A_+) \leq z_-, \tag{6.6}$$

where z_- is the extreme right-hand root of the equation

$$z^n = \|V_-\|_2 \sum_{j=0}^{n-1} \gamma_{n,j} z^{n-j-1} g^j(A).$$

Replace A_+ by $A_- = D + V_-$. Repeating the same arguments, we get

$$sv_A(A_-) \leq z_+, \tag{6.7}$$

where z_+ is the extreme right-hand root of the equation

$$z^n = \|V_+\|_2 \sum_{j=0}^{n-1} \gamma_{n,j} z^{n-j-1} g^j(A).$$

Relations (6.6) and (6.7) imply (6.2). Furthermore, take μ in such a way that $|\mu| = r_s(D)$. Then due to (6.2), there is $\mu_0 \in \sigma(A)$, such that $|\mu_0| \geq r_s(D) - z(\nu)$. Hence, (6.3) follows. Similarly, inequality (6.4) can be proved.

Now take μ in such a way that $Re\ \mu = \alpha(D)$. Due to (6.2) for some $\mu_0 \in \sigma(A)$, $|Re\ \mu_0 - \alpha(D)| \leq z(\nu)$. So, either $Re\ \mu_0 \geq \alpha(D)$, or $Re\ \mu_0 \geq \alpha(D) - z(\nu)$. Thus, inequality (6.5) is also proved. The proof is complete. $\square$

Again put

$$w_n = \sum_{j=0}^{n-1} \gamma_{n,j}.$$

Setting $z = g(A)y$ in (6.1) and applying Lemma 1.6.1, we obtain the estimate $z(\nu) \leq \delta_n(A)$, where

$$\delta_n(A) = \begin{cases} \nu(A)w_n & \text{if } \nu(A)w_n \geq g(A), \\ g^{1-1/n}(A)[\nu(A)w_n]^{1/n} & \text{if } \nu(A)w_n \leq g(A) \end{cases}.$$

Now Theorem 4.6.1 ensures the following result.

Corollary 4.6.2   *For a matrix* $A = (a_{jk})_{j,k=1}^n$, *the inequalities*

$$r_s(A) \geq \max_{k=1,\dots,n} |a_{kk}| - \delta_n(A), \tag{6.8}$$

$$r_l(A) \leq \min_{k=1,\dots,n} |a_{kk}| + \delta_n(A),$$

$$\alpha(A) \geq \max_{k=1,\dots,n} Re\ a_{kk} - \delta_n(A) \tag{6.9}$$

are valid.

4.7   Additional Bounds for Eigenvalues

In the present section instead of the Euclidean norm we use the norm $\|.\|_\infty$. Besides, we derive additional bounds for eigenvalues.

Let A be an $n \times n$-matrix with entries a_{jk} $(j, k = 1, ..., n)$, again. Denote

$$q_{up} = \max_{j=2,...,n} \sum_{k=1}^{j-1} |a_{jk}|, \; q_{low} = \max_{j=1,...,n-1} \sum_{j+1}^{n} |a_{jk}|.$$

Recall that $\tilde{v}_k, \tilde{w}_k, m_{up}(A)$ and $m_{low}(A)$ are defined in Section 3.4. Without lossing the generality, assume that

$$min \; \{q_{low}m_{up}(A), \; q_{up}m_{low}(A)\} \leq 1. \tag{7.1}$$

Theorem 4.7.1 *Under condition (1.1), all the eigenvalues of A lie in the union of the discs*

$$\{\lambda \in \mathbf{C} : |\lambda - a_{kk}| \leq \delta_\infty(A)\}, \; k = 1, ..., n,$$

where

$$\delta_\infty(A) := \sqrt[n]{min\{q_{low}m_{up}(A), \; q_{up}m_{low}(A)\}}.$$

The proof of this theorem is presented in the next section.

If A is a triangular matrix, then Theorem 4.7.1 gives the exact relation

$$\sigma(A) = \{a_{kk}, \; k = 1, ..., n\}.$$

Moreover, we have

Corollary 4.7.2 *Under condition (1.1), the spectral radius $r_s(A)$ of A satisfies the inequality*

$$r_s(A) \leq \max_{k=1,...,n} |a_{kk}| + \delta_\infty(A).$$

Furthermore, consider the quantity

$$s(A) \equiv \max_{i,j} |\lambda_i(A) - \lambda_j(A)|.$$

According to Theorem 4.7.1, for arbitrary $\lambda_i(A), \lambda_j(A)$, there are a_{kk}, a_{mm}, such that

$$|\lambda_i(A) - a_{kk}| \leq \delta_\infty(A), \; |\lambda_j(A) - a_{mm}| \leq \delta_\infty(A).$$

Consequently,

$$|\lambda_i(A) - \lambda_j(A)| \leq |\lambda_i(A) - a_{kk}| + |\lambda_j(A) - a_{mm}| + s(D)$$

$$\leq s(D) + 2\delta_\infty(A) \; (i, j \leq n),$$

where $s(D) = \max_{i,j} |a_{ii} - a_{jj}|$. So we have

Corollary 4.7.3 *Under condition (1.1), $s(A) \leq s(D) + 2\delta_\infty(A)$.*

According to Theorem 4.7.1, $Re\ \lambda_k(A) \leq \max_j Re\ a_{jj} + \delta_\infty(A)$. We thus get

Corollary 4.7.4 *Under (7.1), let $\max_j Re\ a_{jj} + \delta_\infty(A) < 0$. Then A is a Hurwitz matrix.*

These results supplement the well known theorems by Hirsh and Rochrbach (Marcus and Minc, 1964, Sections III.1.3 and Section III.3.1).

4.8 Proof of Theorem 4.7.1

Recall that $V_\pm$ and D are defined in Section 3.1.

Lemma 4.8.1 *The following estimate is true:*

$$\|R_\lambda(D + V_+)\|_\infty \leq \rho^{-1}(D, \lambda) \prod_{k=2}^{n} [1 + \frac{\tilde{v}_k}{|a_{kk} - \lambda|}],$$

where $\rho(D, \lambda) = \min_{k=1,...n} |a_{kk} - \lambda|$.

Proof: Clearly, $R_\lambda(D + V_+) = R_\lambda(D)(I + V_{up}R_\lambda(D))^{-1}$. Due to Lemma 2.10.1,

$$\|(I + V_+R_\lambda(D))^{-1}\|_\infty \leq \prod_{k=2}^{n} [1 + \frac{\tilde{v}_k}{|a_{kk} - \lambda|}],$$

since $V_+R_\lambda(D)$ is nilpotent and its entries are $a_{jk}(a_{kk} - \lambda)^{-1}$. Hence, the required result follows. $\square$

Lemma 4.8.2 *Each eigenvalue μ of the matrix $A = (a_{jk})$ satisfies the inequality*

$$1 \leq q_{low}\rho^{-1}(D, \mu) \prod_{k=2}^{n} (1 + \frac{\tilde{v}_k}{|a_{kk} - \mu|}).$$

Proof: Take $B = V_+ + D$. We have

$$\|A - B\|_\infty = \|V_-\|_\infty = \max_{j=2,...,n} \sum_{k=1}^{j-1} |a_{jk}| = q_{low}.$$

Lemma 4.1.1 and Lemma 4.8.1 imply

$$1 \leq q_{low}\|R_\mu(D + V_+)\| \leq q_{up}\rho^{-1}(D, \mu) \prod_{k=2}^{n} (1 + \frac{\tilde{v}_k}{|a_{kk} - \lambda|})$$

for any $\mu \in \sigma(A)$, as claimed. $\square$

Lemma 4.8.3 *All the eigenvalues of matrix $A = (a_{jk})_{j,k=1}^{n}$ lie in the set*

$$\cup_{j=2}^{n} \overline{\Omega}(a_{kk}, z(A)),$$

where

$$\overline{\Omega}(a_{kk}, z(A)) = \{\lambda \in \mathbf{C} : |\lambda - a_{kk}| \le z(A)\} \quad (k = 1, ..., n)$$

and $z(A)$ is the extreme right-hand (unique positive and simple) root of the algebraic equation

$$z^{n} = q_{low} \prod_{k=2}^{n} (z + \tilde{v}_{k}). \tag{8.1}$$

Proof: Since $|a_{kk} - \lambda| \ge \rho(D, \lambda)$ for all $k = 1, ..., n$, Lemma 4.8.2 implies the inequality

$$1 \le q_{low}\delta^{-1}(D, \mu) \prod_{k=2}^{n} (1 + \tilde{v}_{k})\rho^{-1}(D, \mu)) \tag{8.2}$$

for any eigenvalues μ of A. Dividing the algebraic equation (8.1) by z^{n} we can write down

$$1 = q_{low} z^{-1} \prod_{k=2}^{n} (1 + \tilde{v}_{k} z^{-1}).$$

Comparing this with (8.2) and taking into account that $z(A)$ is the extreme right root of (8.1) we obtain $\rho(A, \lambda) \le z(A)$, as claimed. $\square$

Proof of Theorem 4.7.1: By Lemma 1.6.1,

$$z(A) \le \sqrt[n]{q_{low}m_{up}(A)} \text{ if } q_{low}m_{up}(A) \le 1.$$

Then due to Lemma 4.8.3 all the eigenvalues of A lie in the union of the discs

$$\{\lambda \in \mathbf{C} : |\lambda - a_{kk}| \le \sqrt[n]{q_{low}m_{up}(A)}\}, k = 1, ..., n.$$

Replace A by the adjoint matrix A^{*}, and repeat our arguments. We can assert that all the eigenvalues of A^{*} lie in the union of the discs

$$\{\lambda \in \mathbf{C} : |\lambda - \overline{a}_{kk}| \le \sqrt[n]{q_{up}m_{low}(A)}\}; k = 1, ..., n,$$

provided that $q_{up}m_{low}(A) \le 1$. Since $\sigma(A) = \overline{\sigma}(A^{*})$, we get the required result. $\square$

4.9   Notes

Recall that, for nonnegative matrices, Frobenius has derived the following lower estimate:

$$r_s(A) \geq \tilde{r}(A) \equiv \min_{j=1,\dots,n} \sum_{k=1}^{n} a_{jk}, \tag{9.1}$$

cf. (Marcus and Minc, 1964, Chapter 3, Section 3.1). Relation (6.8) improves estimate (9.1) in the case $a_{jk} \geq 0$ ($j, k = 1, ..., n$), provided that $\max_k a_{kk} - \delta_n(A) > \tilde{r}(A)$. That is, (6.8) is sharper than (9.1) for matrices which are close to triangular ones, since $\delta_n(A) \to 0$, when $V_- \to 0$ or $V_+ \to 0$.

Due to inequality (6.9), matrix A is unstable, when

$$\max_{k=1,\dots,n} Re \, a_{kk} - z(\nu) > 0.$$

The latter result supplements the Rohrbach theorem (Marcus and Minc, 1964, Chapter 3, Section 3.3.3).

A lot of papers and books are devoted to bounds for $m(A, B)$, cf. (Stewart and Sun, 1990), (Bhatia, 1987), (Elsner, 1985), (Phillips, 1990 and 1991) and references therein. One of the recent results belongs to R. Bhatia, L. Elsner and G. Krause (1990). They have proved that

$$m(A, B) \leq 2^{2-1/n} q^{1/n} (\|A\|_2 + \|B\|_2)^{1-1/n}. \tag{9.2}$$

In the paper (Farid, 1992) that inequality was improved in the case when both A and B are normal matrices with spectra on two intersecting lines. Corollary 4.4.6 improves (9.2), if A is close to a triangular matrix.

The contents of Sections 4.2 and 4.3 are taken from (Gil', 1997) while the results of Sections 4.4 and 4.5 are adapted from (Gil', 2001). The material in Section 4.6 is taken from (Gil', 1995).

About other perturbation results see, for instance, the books (Kato, 1966) and (Baumgartel, 1985).

References

[1] Baumgartel, H. (1985). *Analytic Perturbation Theory for Matrices and Operators*. Operator Theory, Advances and Appl., 52. Birkhauser Verlag, Basel, Boston, Stuttgart

[2] Bhatia, R. (1987). *Perturbation Bounds for Matrix Eigenvalues*, Pitman Res. Notes Math. 162, Longman Scientific and Technical, Essex, U.K.

[3] Bhatia R., Elsner, L. and Krause, G. (1990). Bounds for variation of the roots of a polynomial and the eigenvalues of a matrix. *Linear Algebra and Appl.* **142**, 195-209

[4] Elsner, L. (1985). On optimal bound for the spectral variation of two matrices. *Linear Algebra and Appl.* **71**, 77-80.

[5] Farid, F. 0. (1992). The spectral variation for two matrices with spectra on two intersecting lines, *Linear Algebra and Appl.* **177**: 251-273.

[6] Gil', M.I. (1995). *Norm Estimations for Operator-Valued Functions and Applications*. Marcel Dekker, Inc. New York.

[7] Gil', M.I. (1997). A nonsingularity criterion for matrices, *Linear Algebra Appl.* **253**, 79-87.

[8] Gil', M.I. (2001). Invertibility and positive invertibility of matrices, *Linear Algebra and Appl.*, **327**, 95-104

[9] Kato, T. (1966). *Perturbation Theory for Linear Operators*, Springer-Verlag. New York.

[10] Marcus M. and Minc, H. (1964). *A Survey of Matrix Theory and Matrix Inequalities*, Allyn and Bacon, Boston.

[11] Phillips, D. (1990). Improving spectral-variation bound with Chebyshev polynomials, *Linear Algebra Appl.*, **133**, 165-173

[12] Phillips, D. (1991). Resolvent bounds and spectral variation, *Linear Algebra Appl.*, **149**, 35-40

[13] Stewart, G. W. and Ji-guang Sun, (1990). *Matrix Perturbation Theory*, Academic Press, New York.

5. Block Matrices and π-Triangular Matrices

The present chapter is devoted to block matrices. In particular, we derive the invertibility conditions, which supplement the generalized Hadamard criterion and some other well-known results for block matrices.

5.1 Invertibility of Block Matrices

Let an $n \times n$-matrix $A = (a_{jk})_{j,k=1}^n$ be partitioned in the following manner:

$$A = \begin{pmatrix} A_{11} & A_{12} & \dots & A_{1m} \\ A_{21} & A_{22} & \dots & A_{2m} \\ . & \dots & . & . \\ A_{m1} & A_{m2} & \dots & A_{mm} \end{pmatrix}, \tag{1.1}$$

where $m < n$, A_{jk} are matrices. Again $I = I_n$ is the unit operator in $\mathbf{C}^n$ and

$$\|A\|_\infty = \max_{j=1,\dots,n} \sum_{k=1}^n |a_{jk}|.$$

Let the diagonal blocks A_{kk} be invertible. Denote

$$v_k^{up} = \max_{j=1,2,\dots,k-1} \|A_{jk} A_{kk}^{-1}\|_\infty \ (k = 2, \dots, m),$$

and

$$v_k^{low} = \max_{j=k+1,\dots,m} \|A_{jk} A_{kk}^{-1}\|_\infty \ (k = 1, \dots, m-1).$$

Theorem 5.1.1 *Let the diagonal blocks A_{kk} $(k = 1, ..., m)$ be invertible. In addition, with the notations*

$$M_{up} \equiv \prod_{2 \leq k \leq m} (1 + v_k^{up}), \quad M_{low} \equiv \prod_{1 \leq k \leq m-1} (1 + v_k^{low}),$$

let the condition

$$M_{low} M_{up} < M_{low} + M_{up} \tag{1.2}$$

hold. Then matrix A defined by (1.1) is invertible. Moreover,

$$\|A^{-1}\|_\infty \leq \frac{\max_k \|A_{kk}^{-1}\|_\infty M_{low} M_{up}}{M_{low} + M_{up} - M_{low} M_{up}}. \tag{1.3}$$

The proof of this theorem is divided into a series of lemmas, which are presented in Sections 5.2-5.5.

Theorem 5.1.1 is exact in the following sense. Let the matrix A in (1.1) be upper block triangular and A_{kk} be invertible. Then $M_{low} = 1$ and condition (1.2) takes the form $M_{up} < 1 + M_{up}$. Thus, due to Theorem 5.1.1, A is invertible. We have the same result if the matrix in (1.1) is lower block triangular.

Consider a block matrix with $m = 2$:

$$A = \begin{pmatrix} A_{11} & A_{12} \\ A_{21} & A_{22} \end{pmatrix}. \tag{1.4}$$

Then

$$v_2^{up} = \|A_{12} A_{22}^{-1}\|_\infty, \ v_1^{low} = \|A_{21} A_{11}^{-1}\|_\infty.$$

In addition,

$$M_{up} = 1 + v_2^{up} \text{ and } M_{low} = 1 + v_1^{low}.$$

Assume that

$$(1 + v_2^{up})(1 + v_1^{low}) < 2 + v_2^{up} + v_1^{low},$$

or, equivalently, that

$$\|A_{12} A_{22}^{-1}\|_\infty \|A_{21} A_{11}^{-1}\|_\infty < 1. \tag{1.5}$$

Then due to Theorem 5.1.1, the matrix in (1.4) is invertible. Moreover,

$$\|A^{-1}\|_\infty \leq \frac{\max_{k=1,2} \|A_{kk}^{-1}\|_\infty M_{low} M_{up}}{M_{low} + M_{up} - M_{low} M_{up}}.$$

Now assume that m is even: $m = 2m_0$ with a natural m_0, and A_{jk} are 2×2 matrices:

$$A_{jk} = \begin{pmatrix} a_{2j-1,2k-1} & a_{2j-1,2k} \\ a_{2j-1,2k} & a_{2j,2k} \end{pmatrix}$$

with $j, k = 1, ..., m_0$. Take into account that

$$A_{kk}^{-1} = d_k^{-1} \begin{pmatrix} a_{2k,2k} & -a_{2k-1,2k} \\ -a_{2k,2k-1} & a_{2k-1,2k-1} \end{pmatrix}$$

with

$$d_k = a_{2k-1,2k-1} a_{2k,2k} - a_{2k,2k-1} a_{2k-1,2k}.$$

Thus, the quantities v_k^{up}, v_k^{low} are simple to calculate. Now relation (1.2) yields the invertibility, and (1.3) gives the estimate for the inverse matrix.

In Section 5.4 below, we show that the matrix in (1.4) is invertible provided

$$\|A_{12} A_{22}^{-1} A_{21} A_{11}^{-1}\| < 1 \qquad (1.6)$$

with an arbitrary matrix norm $\|.\|$ in $\mathbf{C}^n$.

5.2   π-Triangular Matrices

Let $B(\mathbf{C}^n)$ be the set of all linear operators in $\mathbf{C}^n$. In what follows

$$\pi = \{P_k, \ k = 0, ..., m \le n\}$$

is *a chain of orthogonal projectors* P_k *in* $\mathbf{C}^n$, *such that*

$$0 = P_0 \subset P_1 \subset .... \subset P_m = I_n.$$

The relation $P_{k-1} \subset P_k$ means that

$$P_{k-1}\mathbf{C}^n \subset P_k \mathbf{C}^n \ \ (k = 1, ..., m).$$

Let operators $A, D, V \in B(\mathbf{C}^n)$ satisfy the relations

$$AP_k = P_k A P_k \ (k = 1, ..., m), \qquad (2.1)$$

$$DP_k = P_k D \ (k = 1, ..., m), \qquad (2.2)$$

$$VP_k = P_{k-1} V P_k \ (k = 2, ..., m); \ VP_1 = 0. \qquad (2.3)$$

Then A, D and V *will be called a π-triangular operator, a π- diagonal one and a π-nilpotent one*, respectively

Since

$$V^m = V^m P_m = V^{m-1} P_{m-1} V = V^{m-2} P_{m-2} V P_{m-1} V =$$

$$... = V P_1 ... V P_{m-2} V P_{m-1} V,$$

we have

$$V^m = 0. \qquad (2.4)$$

That is, every π-nilpotent operator is a nilpotent operator. Denote

$$\Delta P_k = P_k - P_{k-1}; \ V_k = V \Delta P_k \ (k = 1, ..., m).$$

Lemma 5.2.1 *Let A be π-triangular. Then*

$$A = D + V, \tag{2.5}$$

where V is a π-nilpotent operator and D is a π-diagonal one.

Proof: Clearly,

$$A = \sum_{j=1}^{m} \Delta P_j A \sum_{k=1}^{m} \Delta P_k = \sum_{k=1}^{m} \sum_{j=1}^{k} \Delta P_j A \Delta P_k = \sum_{k=1}^{m} P_k A \Delta P_k.$$

Hence (2.5) is valid with

$$D = \sum_{k=1}^{m} \Delta P_k A \Delta P_k, \tag{2.6}$$

and

$$V = \sum_{k=2}^{m} P_{k-1} A \Delta P_k = \sum_{k=2}^{m} V_k. \tag{2.7}$$

$\square$

Definition 5.2.2 *Let $A \in B(\mathbf{C}^n)$ be a π-triangular operator and suppose (2.5) holds. Then the π-diagonal operator D and π-nilpotent operator V will be called the π-diagonal part and π-nilpotent part of A, respectively.*

Lemma 5.2.3 *Let $\pi = \{P_k\}_{k=1}^{m}$ be a chain of orthogonal projectors in $\mathbf{C}^n$. If $\tilde{V}$ is a π-nilpotent operator, and A is a π-triangular one, then both operators $A\tilde{V}$ and $\tilde{V}A$ are π-nilpotent ones.*

Proof: By (2.1) and (2.3) we get

$$\tilde{V}AP_k = \tilde{V}P_k A P_k = P_{k-1} \tilde{V} P_k A P_k = P_{k-1} \tilde{V} A P_k, \quad k = 1, ..., m.$$

That is, $\tilde{V}A$ is indeed a π-nilpotent operator. Similarly we can prove that $A\tilde{V}$ is a π-nilpotent operator. $\square$

Lemma 5.2.4 *Let $A \in B(\mathbf{C}^n)$ be a π- triangular operator and D be its π-diagonal part. Then the spectrum of A coincides with the spectrum of D.*

Proof: Due to (2.5)

$$R_\lambda(A) = (A - \lambda I)^{-1} = (D + V - \lambda I)^{-1} = R_\lambda(D)(I + V R_\lambda(D))^{-1}. \tag{2.8}$$

According to Lemma 5.2.3, $V R_\lambda(D)$ is π- nilpotent if λ is not an eigenvalue of D. Therefore,

$$R_\lambda(A) = R_\lambda(D)(I + V R_\lambda(D))^{-1} = \sum_{k=0}^{m-1} R_\lambda(D)(-V R_\lambda(D))^k.$$

This relation implies that λ is a regular point of A if it is a regular point of D.

Conversely, let $\lambda \notin \sigma(A)$. Due to (2.5)

$$R_\lambda(D) = (D - \lambda I)^{-1} = (A - V - \lambda I)^{-1} =$$

$$R_\lambda(A)(I - VR_\lambda(A))^{-1}.$$

According to Lemma 5.2.3, $VR_\lambda(A)$ is π- nilpotent. Therefore,

$$R_\lambda(D) = \sum_{k=0}^{m-1} R_\lambda(A)(VR_\lambda(A))^k.$$

So $\lambda \notin \sigma(D)$, provided let $\lambda \notin \sigma(A)$. The lemma is proved. $\square$

5.3 Multiplicative Representation of Resolvents of π-Triangular Operators

For $X_1, X_2, ..., X_m \in B(\mathbf{C}^n)$ and $j < m$, again denote

$$\overset{\rightarrow}{\prod_{j \leq k \leq m}} X_k \equiv X_j X_{j+1}...X_m.$$

Lemma 5.3.1 *Let* $\pi = \{P_k\}_{k=1}^m$ *be a chain of orthogonal projectors in* $\mathbf{C}^n$ *($m \leq n$), V be a π-nilpotent operator. Then*

$$(I - V)^{-1} = \overset{\rightarrow}{\prod_{2 \leq k \leq m}} (I + V\Delta P_k). \tag{3.1}$$

Proof: According to (2.4)

$$(I - V)^{-1} = \sum_{k=0}^{m-1} V^k. \tag{3.2}$$

On the other hand,

$$\overset{\rightarrow}{\prod_{2 \leq k \leq m}} (I + V\Delta P_k) = I + \sum_{k=2}^{m} V_k + \sum_{2 \leq k_1 < k_2 \leq m} V_{k_1} V_{k_2}$$

$$+ ... + V_2 V_3 ... V_m.$$

Here, as above, $V_k = V\Delta P_k$. However,

$$\sum_{2 \leq k_1 < k_2 \leq m} V_{k_1} V_{k_2} = V \sum_{2 \leq k_1 < k_2 \leq m} \Delta P_{k_1} V \Delta P_{k_2} =$$

$$V \sum_{3 \leq k_2 \leq m} P_{k_1-1} V \Delta P_{k_2} = V^2 \sum_{3 \leq k_2 \leq m} \Delta P_{k_2} = V^2.$$

Similarly,

$$\sum_{2 \leq k_1 < k_3 \ldots < k_j \leq m} V_{k_1} V_{k_2} \ldots V_{k_j} = V^j$$

for $j < m$. Thus from (3.2) follows (3.1). This is the desired conclusion. $\square$.

Theorem 5.3.2 *For any π- triangular operator A and a regular $\lambda \in \mathbf{C}$*

$$R_\lambda(A) = (D - \lambda I)^{-1} \overset{\rightarrow}{\prod_{2 \leq k \leq m}} (I - V \Delta P_k (D - \lambda I)^{-1} \Delta P_k),$$

where D and V are the π-diagonal and π-nilpotent parts of A, respectively.

Proof: Due to Lemma 5.2.3, $V R_\lambda(D)$ is π-nilpotent. Now Lemma 5.3.1 gives

$$(I + V R_\lambda(D))^{-1} = \overset{\rightarrow}{\prod_{2 \leq k \leq m}} (I - V R_\lambda(D) \Delta P_k).$$

But $R_\lambda(D) \Delta P_k = \Delta P_k R_\lambda(D)$. This proves the result. $\square$

5.4   Invertibility with Respect to a Chain of Projectors

Again, let

$$\pi = \{P_k, \ k = 0, \ldots, m\}$$

be a chain of orthogonal projectors P_k. Denote the chain of the complementary projectors by $\tilde{\pi}$:

$$\tilde{P}_k = I_n - P_{m-k} \text{ and } \tilde{\pi} = \{I_n - P_{m-k}, \ k = 0, \ldots, m\}.$$

In the present section, V is a π-nilpotent operator, D is a π-diagonal one, and W is a $\tilde{\pi}$-nilpotent operator.

Lemma 5.4.1 *Any operator $A \in B(\mathbf{C}^n)$ admits the representation*

$$A = D + V + W. \tag{4.1}$$

Proof: Clearly,

$$A = \sum_{j=1}^{m} \Delta P_j A \sum_{k=1}^{m} \Delta P_k.$$

Hence, (4.1) holds, where D and V are defined by (2.6) and (2.7), and

$$W = \sum_{k=1}^{m} \sum_{j=k+1}^{m} \Delta P_j A \Delta P_k = \sum_{k=2}^{m} \tilde{P}_{k-1} A \Delta \tilde{P}_k$$

with $\Delta \tilde{P}_k = \tilde{P}_k - \tilde{P}_{k-1}$. $\square$

Let $\|.\|$ be an arbitrary norm in $\mathbf{C}^n$.

Lemma 5.4.2 *Let the π-diagonal matrix D be invertible. In addition, with the notations*

$$V_A \equiv VD^{-1}, W_A \equiv WD^{-1},$$

let the condition

$$\theta := \|V_A(I + V_A)^{-1}(I + W_A)^{-1}W_A\| < 1 \tag{4.2}$$

hold. Then the operator A defined by (4.1) is invertible. Moreover,

$$\|A^{-1}\| \leq \|(I + V_A)^{-1}\| \|D^{-1}(I + W_A)^{-1}\|(1 - \theta)^{-1}. \tag{4.3}$$

Proof: Due to Lemma 5.4.1,

$$A = D + V + W = (I + V_A + W_A)D = [(I + V_A)(I + W_A) - V_A W_A]D.$$

Clearly, V_A and W_A are nilpotent matrices and, consequently, the matrices, $I + V_A$ and $I + W_A$ are invertible. Thus,

$$A = (I + V_A)(I - B_A)(I + W_A)D, \tag{4.4}$$

where

$$B_A = (I + V_A)^{-1}V_A W_A(I + W_A)^{-1}.$$

Condition (4.2) yields

$$\|(I - B_A)^{-1}\| \leq (1 - \theta)^{-1}.$$

Therefore, (4.2) provides the invertibility of A. Moreover, according to (4.4), inequality (4.3) is valid. $\square$

Corollary 5.4.3 *Under condition (1.6), the matrix in (1.4) is invertible.*

Indeed, under the consideration, $V_A^2 = 0, W_A^2 = 0$. So

$$V_A(I + V_A)^{-1} = V_A \text{ and } W_A(I + W_A)^{-1} = W_A.$$

Hence $\theta = \|V_A W_A\|$. Now Lemma 5.4.2 yields the required result.
   Furthermore, Lemmas 5.3.1 and 5.2.3 yield the following:

Lemma 5.4.4 *The inequalities*

$$\|(I + V_A)^{-1}\| \leq M(V_A) := \prod_{2 \leq k \leq m} (1 + \|VD^{-1}\Delta P_k\|)$$

and

$$\|(I + W_A)^{-1}\| \leq M(W_A) := \prod_{1 \leq k \leq m-1} (1 + \|WD^{-1}\Delta P_k\|)$$

are valid.

5.5   Proof of Theorem 5.1.1

In the present section, D, V and W are the diagonal, upper diagonal and lower diagonal parts of the matrix in (1.1), respectively, that is,

$$D = \begin{pmatrix} A_{11} & 0 & \cdots & 0 \\ 0 & A_{22} & \cdots & 0 \\ \cdot & \cdots & \cdot & \cdot \\ 0 & 0 & \cdots & A_{mm} \end{pmatrix}, V = \begin{pmatrix} 0 & A_{12} & \cdots & A_{1m} \\ 0 & 0 & \cdots & A_{2m} \\ \cdot & \cdots & \cdot & \cdot \\ 0 & 0 & \cdots & 0 \end{pmatrix}, \quad (5.1)$$

and

$$W = \begin{pmatrix} 0 & 0 & \cdots & 0 \\ A_{21} & 0 & \cdots & 0 \\ \cdot & \cdots & \cdot & \cdot \\ A_{m1} & A_{m2} & \cdots & 0 \end{pmatrix}. \quad (5.2)$$

Recall that $V_A \equiv VD^{-1}, W_A \equiv WD^{-1}$.

Lemma 5.5.1 *Let D be invertible. Then the inequalities*

$$\|(I + V_A)^{-1}\|_\infty \leq M_{up} \quad (5.3)$$

and

$$\|(I + W_A)^{-1}\|_\infty \leq M_{low} \quad (5.4)$$

are valid.

Proof: Let $\hat{\pi} = \{\hat{P}_k, \ k = 1, .., m\}$ be the chain of the projectors onto the standard basis. That is, for an $h = (h_k) \in \mathbf{C}^n$,

$$\hat{P}_k h = (h_1, ..., h_{\nu_k}, ..., 0),$$

where $\nu_k = \ dim \ \hat{P}_k$. Then according to (5.1) D and V are $\hat{\pi}$-diagonal and $\hat{\pi}$-nilpotent operators, respectively. Moreover,

$$VD^{-1}\Delta\hat{P}_k = \ V_k A_{kk}^{-1}\Delta\hat{P}_k = \sum_{j=1}^{k-1} A_{jk}A_{kk}^{-1}\Delta\hat{P}_k.$$

But, clearly,

$$\|\sum_{j=1}^{k-1} A_{jk} A_{kk}^{-1} \Delta \hat{P}_k\|_\infty = \max_j \|A_{jk} A_{kk}^{-1}\|_\infty = v_k^{up}.$$

Therefore inequality (5.3) is due to the previous lemma. Inequality (5.4) can be similarly proved. $\square$

Lemma 5.5.2 *Let D be invertible. Then the inequalities*

$$\|V_A(I + V_A)^{-1}\|_\infty \le M_{up} - 1 \tag{5.5}$$

and

$$\|W_A(I + W_A)^{-1}\|_\infty \le M_{low} - 1 \tag{5.6}$$

are valid.

Proof: Let $B = (b_{jk})_{k=1}^n$ be a positive matrix with the property

$$Bh \ge h \tag{5.7}$$

for any nonnegative $h \in \mathbf{C}^n$. Then

$$\|B - I\|_\infty = \max_{j=1,\dots,n} [\sum_{k=1}^n b_{jk} - \delta_{jk}] = \|B\|_\infty - 1. \tag{5.8}$$

Here δ_{jk} is the Kronecker symbol. Furthermore, since V_A is nilpotent,

$$\|V_A(I + V_A)^{-1}\|_\infty \le \|\sum_{k=1}^{n-1} |V_A|^k\|_\infty = \|(I - |V_A|)^{-1} - I\|_\infty,$$

where $|V_A|$ is the matrix whose entries are the absolute values of the entries of V_A. Moreover,

$$\sum_{k=0}^{n-1} |V_A|^k h \ge h$$

for any nonnegative $h \in \mathbf{C}^n$. So according to (5.7) and (5.8)

$$\|V_A(I + V_A)^{-1}\|_\infty \le \|(I - |V_A|)^{-1}\|_\infty - 1.$$

Since

$$\||V_A|\Delta \hat{P}_k\|_\infty = \|V_A \Delta \hat{P}_k\|_\infty,$$

inequality (5.3) with $-|V_A|$ instead of V_A yields inequality (5.5). Inequality (5.6) can be proved similarly. $\square$

   The assertion of Theorem 5.1.1 follows from Lemmas 5.4.2, 5.5.1 and 5.5.2.

5.6 Notes

Many books and papers are devoted to the invertibility of block matrices, cf. (Feingold and Varga, 1962), (Fiedler, 1960), (Gantmacher, 1967), (Ostrowski, 1961), etc. In these works, the Hadamard theorem is mainly generalized to block matrices. Note that the generalized Hadamard theorem does not assert that a block triangular matrix with nonsingular diagonal blocks is invertible. But it is not hard to check that such a matrix is always invertible.

Theorem 5.1.1 gives us the invertibility conditions which improve well-known results for matrices that are "close" to block triangular matrices. Moreover, we derive an estimate for the norm of the inverse matrices.

Chapter 5 is based on the paper (Gil', 2002).

References

[1] Feingold D.G. and Varga, R.S. (1962). Block diagonally dominant matrices and generalization of the Gershgorin circle theorem. *Pacific J. of Mathematics*, **12**, 1241-1250.

[2] Fiedler, M. (1960) Some estimates of spectra of matrices. *Symposium of the Numerical Treatment of Ordinary Differential Equations, Integral and Integro- Differential Equations*, Birkháuser Verlag, Rome, 33-36.

[3] Gantmacher, F. R. (1967). *Theory of Matrices*, Nauka, Moscow (in Russian).

[4] Gil', M.I. (2002). Invertibility and spectrum localization of nonselfadjoint operators, *Adv. Appl. Mathematics*, **28**, 40-58.

[5] Horn, R. A. and Johnson, C. R. (1991). *Topics in Matrix Analysis* , Cambridge University Press, Cambridge.

[6] Ostrowski, A.M. (1961). On some metrical properties of operator matrices and matrices partitioned into blocks. *J. Math. Ann. Appl.*, **2**, 161-209.

6. Norm Estimates for Functions of Compact Operators in a Hilbert Space

The present chapter contains the estimates for the norms of the resolvents and analytic functions of Hilbert-Schmidt operators and resolvents of von Neumann-Schatten operators.

6.1 Bounded Operators in a Hilbert Space

In this section we recall very briefly some basic notions of the theory of operators in a Hilbert space. More details can be found in any textbook on Hilbert spaces (e.g. (Ahiezer and Glazman, 1981), (Dunford and Schwartz, 1963)).

In the sequel H denotes a separable Hilbert space with a scalar product $(.,.)$ and the norm

$$\|h\| = \sqrt{(h,h)} \ \ (h \in H).$$

A sequence $\{h_n\}$ of elements of H converges *strongly* (in the norm) to $h \in H$ if $\|h_n - h\| \to 0$ as $n \to \infty$. Any separable Hilbert space possesses an orthonormal basis. This means that there is a sequence $\{e_k \in H\}$, such that

$$(e_k, e_j) = 0 \text{ if } j \neq k \text{ and } (e_k, e_k) = 1 \ \ (j, k = 1, 2, ...)$$

and any $h \in H$ can be represented as

$$h = \sum_{k=1}^{\infty} c_k e_k$$

with $c_k = (h, e_k)$ $(k = 1, 2, \ldots)$. In addition, this series strongly converges. If the closed linear span of vectors $\{v_k \in H\}_{k=1}^{\infty}$ coincides with H, then the set of these vectors is said to be complete in H.

A linear operator A acting in H is called a bounded one, if there is a constant a such that

$$\|Ah\| \leq a\|h\| \text{ for all } h \in H.$$

The quantity

$$\|A\| = \sup_{h \in H} \frac{\|Ah\|}{\|h\|}$$

is called the norm of A. A sequence $\{A_n\}$ of bounded linear operators *converges strongly* to an operator A, if the sequence of elements $\{A_n h\}$ strongly converges to Ah for every $h \in H$. $\{A_n\}$ converges *in the uniform operator topology* (in the operator norm) to an operator A, if $\|A_n - A\| \to 0$ as $n \to \infty$. A bounded linear operator A^* is called adjoint to A, if

$$(Af, g) = (f, A^* g) \text{ for every } h, g \in H.$$

The relation $\|A\| = \|A^*\|$ is true. A bounded operator A *is a selfadjoint* one, if $A = A^*$. A is *a unitary operator*, if $AA^* = A^* A = I$. Here and below $I \equiv I_H$ is the identity operator in H. A selfadjoint operator A is positive (negative) definite, if

$$(Ah, h) \geq 0 \quad ((Ah, h) \leq 0) \text{ for every } h \in H.$$

A selfadjoint operator A is strongly positive (strongly negative) definite, if there is a constant $c > 0$, such that

$$(Ah, h) \geq c \, (h, h) \quad ((Ah, h) < -c \, (h, h)) \text{ for every } h \in H.$$

A bounded linear operator satisfying the relation $AA^* = A^* A$ is called *a normal operator*. It is clear that unitary and selfadjoint operators are examples of normal ones. The operator $B \equiv A^{-1}$ is the inverse one to A, if $AB = BA = I$. An operator P is called *a projector* if $P^2 = P$. If, in addition, $P^* = P$, then it is called *an orthogonal projector (an orthoprojector)*.

A point λ of the complex plane is said to be a regular point of an operator A, if the operator $R_\lambda(A) \equiv (A - I\lambda)^{-1}$ (the resolvent) exists and is bounded. The complement of all regular points of A in the complex plane is the *spectrum* of A. The spectrum of A is denoted by $\sigma(A)$. The spectrum of a selfadjoint operator is real, the spectrum of a unitary operator lies on the unit circle.

The quantity

$$r_s(A) = sup_{s \in \sigma(A)}|s|$$

is the *spectral radius* of A. *An operator V is called a quasinilpotent one, if its spectrum consists of zero, only.* If there is a nontrivial solution e of the equation

$$Ae = \lambda(A)e,$$

where $\lambda(A)$ is a number, then this number is called an eigenvalue of operator A, and $e \in H$ is an eigenvector corresponding to $\lambda(A)$. Any eigenvalue is a point of the spectrum. An eigenvalue $\lambda(A)$ has the (algebraic) multiplicity $r \leq \infty$ if

$$dim(\cup_{k=1}^{\infty} ker(A - \lambda(A)I)^k) = r.$$

In the sequel $\lambda_k(A)$, $k = 1, 2, ...$ are the eigenvalues of A repeated according to their multiplicities.

A vector v satisfying $(A - \lambda(A)I)^n v = 0$ for a natural n, is a root vector of operator A corresponding to $\lambda(A)$.

6.2 Compact Operators in a Hilbert Space

All the results, presented in this section can be found, for instance, in (Gohberg and Krein, 1969, Chapters 2 and 3). *The set of all linear completely continuous (compact) operators in H is defined by C_∞.*

Recall that the spectrum of an operator from C_∞ is either finite, or the sequence of the eigenvalues of A converges to zero, any nonzero eigenvalue has the finite multiplicity. Moreover, any normal operator $A \in C_\infty$ can be represented in the form

$$A = \sum_{k=1}^{\infty} \lambda_k(A)E_k, \qquad (2.1)$$

where E_k are eigenprojectors of A, i.e. the projectors defined by $E_k h = (h, d_k)d_k$ for all $h \in H$. Here d_k are the normal eigenvectors of A. Recall that eigenvectors of normal operators are mutually orthogonal. A completely continuous positive definite selfadjoint operator has non-negative eigenvalues, only. Let $A \in C_\infty$ be positive definite and represented by (2.1). Then we write

$$A^\beta := \sum_{k=1}^{\infty} \lambda_k^\beta(A)E_k \ \ (\beta > 0).$$

A completely continuous quasinilpotent operator is called a Volterra operator.

Let $\{e_k\}$ be an orthogonal normal basis in H, and the series

$$\sum_{k=1}^{\infty}(Ae_k, e_k) \ \ (A \in C_\infty)$$

converges. Then the sum of this series is called *the trace of A:*

$$Trace\ A = Tr\ A = \sum_{k=1}^{\infty}(Ae_k, e_k).$$

An operator A satisfying the condition

$$Tr\ (A^*A)^{1/2} < \infty$$

is called *a nuclear operator.* An operator A, satisfying the relation

$$Tr\ (A^*A) < \infty$$

is said to be *a Hilbert-Schmidt operator.*

The eigenvalues $\lambda_k((A^*A)^{1/2})$ $(k = 1, 2, ...)$ of the operator $(A^*A)^{1/2}$ are called *the singular numbers* (*s*-numbers) of A and are denoted by $s_k(A)$. That is,

$$s_k(A) \equiv \lambda_k((A^*A)^{1/2})\ (k = 1, 2, ...).$$

Enumerate singular numbers of A taking into account their multiplicity and in decreasing order. The set of completely continuous operators acting in a Hilbert space and satisfying the condition

$$N_p(A) := [\ \sum_{k=1}^{\infty} s_k^p(A)\ ]^{1/p} < \infty,$$

for some $p \geq 1$, is called *the von Neumann - Schatten ideal and is denoted by C_p.* $N_p(.)$ *is called the norm of the ideal C_p.* It is not hard to show that

$$N_p(A) = \sqrt[p]{Tr\ (AA^*)^{p/2}}.$$

Thus, C_1 is the ideal of nuclear operators (*the Trace class*) and C_2 is the ideal of Hilbert-Schmidt operators. $N_2(A)$ is called the *Hilbert-Schmidt norm.* Sometimes we will omit index 2 of the Hilbert-Schmidt norm, i.e.

$$N(A) := N_2(A) = \sqrt{Tr\ (A^*A)}.$$

For any orthogonal normal basis $\{e_k\}$ we can write

$$N_2(A) = (\sum_{k=1}^{\infty} \|Ae_k\|^2)^{1/2}.$$

This equality is equivalent to the following one:

$$N_2(A) = (\sum_{j,k=1}^{\infty} |a_{jk}|^2)^{1/2}, \tag{2.2}$$

where $a_{jk} = (Ae_k, e_j)$ $(j, k = 1, 2, \ldots)$ are entries of a Hilbert-Schmidt operator A in basis $\{e_k\}$.

For all $p \geq 1$, the following propositions are true (the proofs can be found in the books (Gohberg and Krein, 1969, Section 3.7), and (Pietsch, 1988)):

If $A \in C_p$, then also $A^* \in C_p$. If $A \in C_p$ and B is a bounded linear operator, then both AB and BA belong to C_p. Moreover,

$$N_p(AB) \leq N_p(A)\|B\| \text{ and } N_p(BA) \leq N_p(A)\|B\|.$$

In addition, the inequality

$$\sum_{j=1}^{n} |\lambda_j(A)|^p \leq \sum_{j=1}^{n} s_j^p(A) \quad (n = 1, 2, \ldots) \tag{2.3}$$

is valid, cf. (Gohberg and Krein, 1969, Theorem II.3.1).

Lemma 6.2.1 *If $A \in C_p$ and $B \in C_q$ $(1 < p, q < \infty)$, then $AB \in C_s$ with*

$$\frac{1}{s} = \frac{1}{p} + \frac{1}{q}.$$

Moreover,

$$N_s(AB) \leq N_p(A)N_q(B). \tag{2.4}$$

For the proof of this lemma see (Gohberg and Krein, 1969, Section III.7). We need also the following result (Lidskij's theorem).

Theorem 6.2.2 *The trace of $A \in C_1$ does not depend on a choice of an orthogonal normal basis and*

$$Tr\ A = \sum_{k=1}^{\infty} \lambda_k(A).$$

The proof of this theorem can be found in (Gohberg and Krein, 1969, Section III.8).

6.3 Triangular Representations of Compact Operators

Let R_0 be a set in the complex plane and let $\epsilon > 0$. By $S(R_0, \epsilon)$ we denote the ϵ-neighborhood of R_0. That is,

$$dist\{R_0, S(R_0, \epsilon)\} \leq \epsilon.$$

Lemma 6.3.1 *Let A be a bounded operator and let $\epsilon > 0$. Then there is a $\delta > 0$, such that, if a bounded operator B satisfies the condition $\|A - B\| \le \delta$, then $\sigma(B)$ lies in $S(\sigma(A), \epsilon)$ and*

$$\|R_\lambda(A) - R_\lambda(B)\| \le \epsilon$$

for any λ, which does not belong to $S(\sigma(A), \epsilon)$.

For the proof of this lemma we refer the reader to the book (Dunford and Schwartz, 1963, p. 585).

Lemma 6.3.2 *Let $V \in C_p$, $p > 1$ be a Volterra operator. Then there is a sequence of nilpotent operators, having finite dimensional ranges and converging to V in the norm $N_p(.)$.*

Proof: Let $T = V - V^*$. Due to the well-known Theorems 22.1 and 16.3 from the book (Brodskii, 1971), for an $\epsilon > 0$, there is a finite chain $\{P_k\}_{k=0}^n$ of orthogonal projectors onto invariant subspaces of V:

$$0 = Range(P_0) \subset Range(P_1) \subset \dots \subset Range(P_n) = H,$$

such that with the notation

$$W_n = \sum_{k=1}^n P_{k-1} T \Delta P_k \quad (\Delta P_k = P_k - P_{k-1}),$$

the inequality $N_p(W_n - V) < \epsilon$ is valid. Furthermore, let $\{e_m^{(k)}\}_{m=1}^\infty$ be an orthonormal basis in $\Delta P_k H$. Put

$$Q_l^{(k)} = \sum_{m=1}^l (., e_m^{(k)}) e_m^{(k)} \quad (k = 1, ..., n; \; l = 1, 2, ...).$$

Clearly, $Q_l^{(k)}$ strongly converge to ΔP_k as $l \to \infty$. Moreover,

$$Q_l^{(k)} \Delta P_k = \Delta P_k Q_l^{(k)} = Q_l^{(k)}.$$

Since,

$$W_n = \sum_{k=1}^n \sum_{j=1}^{k-1} \Delta P_j T \Delta P_k,$$

the operators

$$W_{nl} = \sum_{k=1}^n \sum_{j=1}^{k-1} Q_l^{(j)} T Q_l^{(k)}$$

have finite dimensional ranges and tend to W_n in the norm N_p as $l \to \infty$, since $T \in C_p$. Thus, W_{nl} tend to V in the norm N_p as $l, n \to \infty$. Put

$$L_k^{(l)} = \sum_{j=1}^k Q_l^{(j)} \quad (k = 1, ..., n).$$

Then $L_{k-1}^{(l)} W_{nl} L_k^{(l)} = W_{nl} L_k^{(l)}$. Hence we easily have $W_{nl}^n = 0$. This proves the lemma. $\square$

We recall the following well-known result, cf. (Gohberg and Krein, 1969, Lemma I.4.2).

Lemma 6.3.3 *Let $M \neq H$ be the closed linear span of all the root vectors of an operator $A \in C_\infty$ and let Q_A be the orthogonal projector of H onto $M^\perp$, where $M^\perp$ is the orthogonal complement of M in H. Then $Q_A A Q_A$ is a Volterra operator.*

The previous lemma means that A can be represented by the matrix

$$A = \begin{pmatrix} B_A & A_{12} \\ 0 & V_1 \end{pmatrix} \tag{3.1}$$

acting in $M \oplus M^\perp$. Here $B_A = A(I - Q_A)$, $V_1 = Q_A A Q_A$ is a Volterra operator in $Q_A H$ and $A_{12} = (I - Q_A) A Q_A$.

Theorem 6.3.4 *Let $A \in C_\infty$. Then there are a normal operator D and a Volterra operator V, such that*

$$A = D + V \ \text{and} \ \sigma(D) = \sigma(A). \tag{3.2}$$

Moreover, A, D and V have the same invariant subspaces.

Proof: Let M be the linear closed span of all the root vectors of A, and P_A is the projector of H onto M. So the system of the root vectors of the operator $B_A = A P_A$ is complete in M. Thanks to the well-known Lemma I.3.1 from (Gohberg and Krein, 1969), there is an orthonormal basis (Schur's basis) $\{e_k\}$ in M, such that

$$B_A e_j = A e_j = \lambda_j(B_A) e_j + \sum_{k=1}^{j-1} a_{jk} e_k \ (j = 1, 2, \ldots). \tag{3.3}$$

We have $B_A = D_B + V_B$, where $D_B e_k = \lambda_k(B_A) e_k$, $k = 1, 2, \ldots$ and $V_B = B_A - D_B$ is a quasinilpotent operator. But according to (3.1) $\lambda_k(B_A) = \lambda_k(A)$, since V_1 is a quasinilpotent operator. Moreover D_B and V_B have the same invariant subspaces. Take the following operator matrix acting in $M \oplus M^\perp$:

$$D = \begin{pmatrix} D_B & 0 \\ 0 & 0 \end{pmatrix} \ \text{and} \ V = \begin{pmatrix} V_B & A_{12} \\ 0 & V_1 \end{pmatrix}.$$

Since the diagonal of V contains V_B and V_1 only, $\sigma(V) = \sigma(V_B) \cup \sigma(V_1) = \{0\}$. So V is quasinilpotent and (3.2) is proved. From (3.1) and (3.3) it follows that A, D and V have the same invariant subspace, as claimed. $\square$

Definition 6.3.5 *Equality (3.2) is said to be the triangular representation of A. Besides, D and V will be called the diagonal part and nilpotent part of A, respectively.*

Lemma 6.3.6 *Let $A \in C_p$, $p \geq 1$. Let V be the nilpotent part of A. Then there exists a sequence $\{A_n\}$ of operators, having n-dimensional ranges, such that*

$$\sigma(A_n) \subseteq \sigma(A), \tag{3.4}$$

and

$$\sum_{k=1}^{n} |\lambda(A_n)|^p \to \sum_{k=1}^{\infty} |\lambda(A)|^p \text{ as } n \to \infty. \tag{3.5}$$

Moreover,

$$N_p(A_n - A) \to 0 \text{ and } N_p(V_n - V) \to 0 \text{ as } n \to \infty, \tag{3.6}$$

where V_n are the nilpotent parts of A_n $(n = 1, 2, ...)$.

Proof: Again, let M be the linear closed span of all the root vectors of A, and P_A the projector of H onto M. So the system of root vectors of the operator $B_A = AP_A$ is complete in M. Let D_B and V_B be the nilpotent parts of B_A, respectively. According to (3.3), put

$$P_n = \sum_{k=1}^{n} (., e_k)e_k.$$

Then

$$\sigma(B_A P_n) = \sigma(D_B P_n) = \{\lambda_1(A), ..., \lambda_n(A)\}. \tag{3.7}$$

In addition, $D_B P_n$ and $V_B P_n$ are the diagonal and nilpotent parts of $B_A P_n$, respectively. Due to Lemma 6.3.2, there exists a sequence $\{W_n\}$ of nilpotent operators having n-dimensional ranges and converging in N_p to the operator V_1. Put

$$A_n = \begin{pmatrix} B_A P_n & P_n A_{12} \\ 0 & W_n \end{pmatrix}.$$

Then the diagonal part of A_n is

$$D_n = \begin{pmatrix} D_B P_n & 0 \\ 0 & 0 \end{pmatrix}$$

and the nilpotent part is

$$V_n = \begin{pmatrix} V_B P_n & P_n A_{12} \\ 0 & W_n \end{pmatrix}.$$

So relations (3.6) are valid. According to (3.7), relation (3.5) holds. Moreover $N_p(D_n - D_B) \to 0$. So relation (3.5) is also proved. This finishes the proof. $\square$

6.4 Resolvents of Hilbert-Schmidt Operators

Let A be a Hilbert-Schmidt operator. The following quantity plays a key role in this section:

$$g(A) = [N_2^2(A) - \sum_{k=1}^{\infty} |\lambda_k(A)|^2]^{1/2}, \tag{4.1}$$

where $N_2(A)$ is the Hilbert-Schmidt norm of A, again. Since

$$\sum_{k=1}^{\infty} |\lambda_k(A)|^2 \geq |\sum_{k=1}^{\infty} \lambda_k^2(A)| = |Trace\ A^2|,$$

one can write

$$g^2(A) \leq N_2^2(A) - |Trace\ A^2|. \tag{4.2}$$

If A is a normal Hilbert-Schmidt operator, then $g(A) = 0$, since

$$N_2^2(A) = \sum_{k=1}^{\infty} |\lambda_k(A)|^2$$

in this case. Let $A_I = (A - A^*)/2i$. We will also prove the inequality

$$g^2(A) \leq \frac{N_2^2(A - A^*)}{2} = 2N_2^2(A_I) \tag{4.3}$$

(see Lemma 6.5.2 below). Again put $\rho(A, \lambda) := \inf_{t \in \sigma(A)} |\lambda - t|$.

Theorem 6.4.1 *Let A be a Hilbert-Schmidt operator. Then*

$$\|R_\lambda(A)\| \leq \sum_{k=0}^{\infty} \frac{g^k(A)}{\rho^{k+1}(A, \lambda)\sqrt{k!}} \quad (\lambda \notin \sigma(A)). \tag{4.4}$$

Proof: Due to Lemma 6.3.6 there exists a sequence $\{A_n\}$ of operators, having n-dimension ranges, such that the relations (3.4),

$$N_2(A_n) \to N_2(A) \text{ and } g(A_n) \to g(A) \text{ as } n \to \infty \tag{4.5}$$

are valid. But due to Corollary 2.1.2,

$$\|R_\lambda(A_n)\| \leq \sum_{k=0}^{n-1} \frac{g^k(A_n)}{\rho^{k+1}(A_n, \lambda)\sqrt{k!}} \quad (\lambda \notin \sigma(A_n)).$$

According to (3.4) $\rho(A_n, \lambda) \geq \rho(A, \lambda)$. Now, letting $n \to \infty$ in the latter relation, we arrive at the stated result. $\square$

An additional proof of this theorem can be found in (Gil', 1995, Chapter 2).

Theorem 6.4.1 is precise. Inequality (4.4) becomes the equality

$$\|R_\lambda(A)\| = \rho^{-1}(A, \lambda),$$

if A is a normal operator, since $g(A) = 0$ in this case.

Note that for an arbitrary constant $c > 1$, Schwarz's inequality implies the relations

$$\sum_{k=0}^{\infty} \frac{x^k}{\sqrt{k!}} = \sum_{k=0}^{\infty} \frac{\sqrt{c^k} x^k}{\sqrt{c^k k!}} \le \Big[\sum_{k=0}^{\infty} \frac{c^k x^{2k}}{k!} \sum_{j=0}^{\infty} \frac{1}{c^j} \Big]^{1/2} = \sqrt{\frac{c}{c-1}} \, e^{cx^2/2} \quad (x \ge 0).$$
(4.6)

With $c = 2$, we have

$$\sum_{k=0}^{\infty} \frac{x^k}{\sqrt{k!}} \le \sqrt{2} \, e^{x^2} \quad (x \ge 0).$$

Now Theorem 6.4.1 implies the inequality

$$\|R_\lambda(A)\| \le \frac{a_0}{\rho(A, \lambda)} exp \, \Big[\frac{b_0 \, g^2(A)}{\rho^2(A, \lambda)} \Big] \text{ for all regular } \lambda,$$
(4.7)

where according to (4.6), one can take.

$$a_0 = \sqrt{\frac{c}{c-1}} \text{ and } b_0 = \frac{c}{2} \text{ for any } c > 1. \text{ In particular, } a_0 = \sqrt{2} \text{ and } b_0 = 1.$$
(4.8)

Moreover, letting $n \to \infty$ in Theorem 2.14.1, we get (4.7) with

$$a_0 = e^{1/2} \text{ and } b_0 = 1/2.$$
(4.9)

We thus have proved

Theorem 6.4.2 *Let $A \in C_2$. Then there are nonnegative constants a_0, b_0, such that estimate (4.7) is valid. Moreover, a_0 and b_0 can be taken as in (4.8) or in (4.9).*

In particular, if $V \in C_2$ is a quasinilpotent operator, then

$$\|R_\lambda(V)\| \le \frac{a_0}{|\lambda|} exp \, \Big[\frac{b_0 \, N_2^2(V)}{|\lambda|^2} \Big] \text{ for all } \lambda \ne 0.$$

6.5   Equalities for Eigenvalues of a Hilbert-Schmidt Operator

Lemma 6.5.1 *Let V be a Volterra operator and $V_I \equiv (V - V^*)/2i \in C_2$. Then $V \in C_2$. Moreover, $N_2^2(V) = 2N_2^2(V_I)$.*

Proof: By Theorem 6.2.2 we have $Trace\ V^2 = Trace\ (V^*)^2 = 0$, because V is a Volterra operator. Hence,

$$N_2^2(V - V^*) = Trace\ (V - V^*)^2 = Trace\ (V^2 + VV^* + V^*V + (V^*)^2)$$
$$= Trace\ (VV^* + V^*V) = 2Trace\ (VV^*).$$

We arrive at the result. $\square$

Lemma 6.5.2 *Let $A \in C_2$. Then*

$$N_2^2(A) - \sum_{k=1}^{\infty} |\lambda_k(A)|^2 = 2N_2^2(A_I) - 2\sum_{k=1}^{\infty} |Im\ \lambda_k(A)|^2 = N_2^2(V),$$

where V is the nilpotent part of A.

Proof: Let D be the diagonal part of A. By (3.3), it is simple to check that VD^* is a Volterra operator (see also Lemma 7.3.4). By Theorem 6.2.2,

$$Trace\ VD^* = Trace\ V^*D = 0. \tag{5.1}$$

From the triangular representation (3.2) it follows that

$$Tr\ AA^* = Tr\ (D + V)(D^* + V^*) = Tr\ AA^* = Tr\ (DD^*) + Tr\ (VV^*).$$

Besides, due to (3.2) $\sigma(A) = \sigma(D)$. Thus,

$$N_2^2(D) = \sum_{k=1}^{\infty} |\lambda_k(A)|^2.$$

So the relation

$$N_2^2(V) = N_2^2(A) - \sum_{k=1}^{\infty} |\lambda_k(A)|^2$$

is proved. Furthermore, from the triangular representation (3.2) it follows that

$$-4Tr\ A_I^2 = Tr\ (A - A^*)^2 = Tr\ (D + V - D^* - V^*)^2.$$

Hence, thanks to (5.1), we obtain

$$-4Tr\ A_I^2 = Tr\ (D - D^*)^2 + Tr\ (V - V^*)^2.$$

That is, $N_2^2(A_I) = N_2^2(V_I) + N_2^2(D_I)$, where $V_I = (V - V^*)/2i$ and $D_I = (D - D^*)/2i$. Taking into account Lemma 6.5.1, we arrive at the equality

$$2N_2^2(A_I) - 2N_2^2(D_I) = N_2^2(V).$$

Besides, due to (3.2) $\sigma(A) = \sigma(D)$. Thus,

$$N^2(D_I) = \sum_{k=1}^{\infty} |Im\, \lambda_k(A)|^2,$$

and we arrive at the required result. $\square$

Replace in Lemma 6.5.2, operator A by Ae^{it} and $Ae^{i\tau}$ with real numbers t, τ. Then we get

Corollary 6.5.3 *Let $A \in C_2$. Then*

$$N^2(Ae^{it} - A^*e^{-it}) - \sum_{k=1}^{\infty} |e^{it}\lambda_k(A) - e^{-it}\overline{\lambda}_k(A)|^2 =$$

$$N^2(Ae^{i\tau} - A^*e^{-i\tau}) - \sum_{k=1}^{\infty} |e^{i\tau}\lambda_k(A) - e^{-i\tau}\overline{\lambda}_k(A)|^2 \quad (t, \tau \in \mathbf{R}).$$

In particular, take $t = 0$ and $\tau = \pi/2$. Then due to Corollary 6.5.3,

$$N^2(A_I) - \sum_{k=1}^{\infty} |Im\, \lambda_k(A)|^2 = N^2(A_R) - \sum_{k=1}^{\infty} |Re\, \lambda_k(A)|^2 \qquad (5.2)$$

with $A_R = (A + A^*)/2$.

6.6 Operators Having Hilbert-Schmidt Powers

Assume that for some positive integer $p > 1$,

$$A^p \text{ is a Hilbert-Schmidt operator.} \qquad (6.1)$$

Note that under (6.1) A can, in general, be a noncompact operator. Below in this section we will give a relevant example.

Theorem 6.6.1 *Let (6.1) hold for some integer $p > 1$. Then*

$$\|R_\lambda(A)\| \le \|T_{\lambda,p}\| \sum_{k=0}^{\infty} \frac{g^k(A^p)}{\rho^{k+1}(A^p, \lambda^p)\sqrt{k!}} \quad (\lambda^p \notin \sigma(A^p)), \qquad (6.2)$$

where

$$T_{\lambda,p} = \sum_{k=0}^{p-1} A^k \lambda^{p-k-1}, \qquad (6.3)$$

and

$$\rho(A^p, \lambda^p) = \inf_{t \in \sigma(A)} |t^p - \lambda^p|$$

is the distance between $\sigma(A^p)$ and the point λ^p.

Proof: We use the identity

$$A^p - I\lambda^p = (A - I\lambda) \sum_{k=0}^{p-1} A^k \lambda^{p-k-1} = (A - I\lambda)T_{\lambda,p}.$$

This implies

$$(A - I\lambda)^{-1} = T_{\lambda,p}(A^p - I\lambda^p)^{-1}. \tag{6.4}$$

Thus,

$$\|(A - I\lambda)^{-1}\| \le \|T_{\lambda,p}\| \, \|(A^p - I\lambda^p)^{-1}\|. \tag{6.5}$$

Applying Theorem 6.4.1 to the resolvent $(A^p - I\lambda^p)^{-1} = R_{\lambda^p}(A^p)$, we obtain:

$$\|R_{\lambda^p}(A^p)\| \le \sum_{k=0}^{\infty} \frac{g^k(A^p)}{\rho^{k+1}(A^p, \lambda^p)\sqrt{k!}} \quad (\lambda^p \notin \sigma(A^p)).$$

This and (6.4) complete the proof. $\square$

According to (6.5) Theorem 6.4.2 gives us

Corollary 6.6.2 *Let condition (6.1) hold for some integer $p > 1$. Then*

$$\|R_\lambda(A)\| \le \frac{a_0\|T_{\lambda,p}\|}{\rho(A^p, \lambda^p)} \, exp\, [\frac{b_0 g^2(A^p)}{\rho^2(A^p, \lambda^p)}] \quad (\lambda^p \notin \sigma(A^p)),$$

where constants a_0 and b_0 can be taken from (4.8) or from (4.9).

Example 6.6.3 *Consider a noncompact operator satisfying condition (6.1).*

Let H be an orthogonal sum of Hilbert spaces H_1 and H_2: $H = H_1 \oplus H_2$, and let A be a linear operator defined in H by the formula

$$A = \begin{pmatrix} B_1 & T \\ 0 & B_2 \end{pmatrix},$$

where B_1 and B_2 are bounded linear operators acting in H_1 and H_2, respectively, and a bounded linear operator T maps H_2 into H_1. Evidently A^2 is defined by the matrix

$$A^2 = \begin{pmatrix} B_1^2 & B_1 T + T B_2 \\ 0 & B_2^2 \end{pmatrix}.$$

If $B_1, B_2 \in C_2$ and T is a noncompact one, then $A^2 \in C_2$, while A is a noncompact operator.

6.7 Resolvents of Neumann-Schatten Operators

Let
$$A \in C_{2p} \text{ for some integer } p > 1. \tag{7.1}$$

Then due to (2.4), condition (6.1) holds. So we can directly apply Theorem 6.6.1, but in appropriate situations the following result is more convenient.

Theorem 6.7.1 *Let* $A \in C_{2p}$ $(p = 2, 3, ...)$. *Then*

$$\|R_\lambda(A)\| \leq \sum_{m=0}^{p-1} \sum_{k=0}^{\infty} \frac{(2N_{2p}(A))^{pk+m}}{\rho^{pk+m+1}(A, \lambda)\sqrt{k!}} \quad (\lambda \notin \sigma(A)). \tag{7.2}$$

The proofs of this theorem and the next one are presented in the next section. An additional proof of Theorem 6.7.1 can be found in (Gil', 1995, Section 2.6). Put

$$\theta_j^{(p)} = \frac{1}{\sqrt{[j/p]!}}, \tag{7.3}$$

where $[x]$ means the integer part of a real number x. Now the previous theorem implies

Corollary 6.7.2 *Let* $A \in C_{2p}$ $(p = 2, 3, ...)$. *Then*

$$\|R_\lambda(A)\| \leq \sum_{j=0}^{\infty} \frac{\theta_j^{(p)}(2N_{2p}(A))^j}{\rho^{j+1}(A, \lambda)} \quad (\lambda \notin \sigma(A)). \tag{7.4}$$

Theorem 6.7.3 *Under condition (7.1) there are constants* $a_0, b_0 > 0$, *such that the estimate*

$$\|R_\lambda(A)\| \leq a_0 \sum_{m=0}^{p-1} \frac{(2N_{2p}(A))^m}{\rho^{m+1}(A, \lambda)} \, exp\, [\frac{b_0(2N_{2p}(A))^{2p}}{\rho^{2p}(A, \lambda)}] \quad (\lambda \notin \sigma(A)) \tag{7.5}$$

holds. These constants can be taken as in (4.8) or in (4.9).

Since, condition (7.1) implies $A_I \equiv (A - A^*)/2i \in C_{2p}$, *additional estimates for the resolvent under condition (7.1) are derived in Section 7.9 below.*

6.8 Proofs of Theorems 6.7.1 and 6.7.3

We need the following result.

Lemma 6.8.1 *Let A be a linear operator acting in a Euclidean space $\mathbf{C}^n$ with $n = jp$ and integers $p \geq 1, j > 1$. Then*

$$\|R_\lambda(A)\| \leq \sum_{m=0}^{p-1} \sum_{k=0}^{j} \frac{N_{2p}^{kp+m}(V)}{\rho^{pk+m+1}(A,\lambda)\sqrt{k!}} \quad (\lambda \notin \sigma(A)) \qquad (8.1)$$

where V is the nilpotent part of A, $\|.\|$ is the Euclidean norm.

Proof: Since $A = D + V$, where D is the diagonal part of A,

$$(A - \lambda I)^{-1} = (D + V - \lambda I)^{-1} = (D - \lambda I)^{-1}(I + B_\lambda)^{-1} \qquad (8.2)$$

where $B_\lambda = -VR_\lambda(D)$. By the identity

$$(I - B_\lambda)(I + B_\lambda + \ldots + B_\lambda^{p-1}) = I - B_\lambda^p$$

we have

$$(A - \lambda I)^{-1} = (D + V - \lambda I)^{-1} =$$
$$(D - \lambda I)^{-1}(I + B_\lambda + \ldots + B_\lambda^{p-1})(I - B_\lambda^p)^{-1}. \qquad (8.3)$$

Clearly, B_λ is a nilpotent operator. So $B_\lambda^n = B_\lambda^{pj} = 0$ and

$$(I - B_\lambda^p)^{-1} = \sum_{k=0}^{j} B_\lambda^{kp}.$$

Thus,

$$(I - B_\lambda)^{-1} = (I + B_\lambda + \ldots + B_\lambda^{p-1})(I - B_\lambda^p)^{-1} = \sum_{m=0}^{p-1}\sum_{k=0}^{j} B_\lambda^{kp+m}.$$

Hence,

$$R_\lambda(A) = R_\lambda(D)\sum_{m=0}^{p-1}\sum_{k=0}^{j} B_\lambda^{kp+m}. \qquad (8.4)$$

Taking into account that $\sigma(A) = \sigma(D)$, we can assert that $R_\lambda(D)$ is a bounded operator for all regular points λ of A. But

$$N_2(B_\lambda^p) \leq N_{2p}^p(B_\lambda) \qquad (8.5)$$

(see relation (2.4)). Now let us use Theorem 2.5.1. It gives

$$\|B_\lambda^{pk}\| \leq \gamma_{n,k}N_2^k(B_\lambda^p) \leq \frac{N_2^k(B_\lambda^p)}{\sqrt{k!}}.$$

Thus, (8.5) ensures the estimate

$$\|B_\lambda^{pk}\| \leq \frac{N_{2p}^{kp}(B_\lambda)}{\sqrt{k!}}.$$

Furthermore, it is clear that

$$N_{2p}(B_\lambda) = N_{2p}(VR_\lambda(D)) \leq N_{2p}(V)\|R_\lambda(D)\|.$$

Since D is normal and $\sigma(A) = \sigma(D)$,

$$\|R_\lambda(D)\| = \rho^{-1}(D, \lambda) = \rho^{-1}(A, \lambda). \tag{8.6}$$

Hence,

$$N_{2p}(B_\lambda) \leq \frac{N_{2p}(V)}{\rho(A, \lambda)}. \tag{8.7}$$

Thus,

$$\|B_\lambda^{pk}\| \leq \frac{N_{2p}^{kp}(V)}{\rho^{kp}(A, \lambda)\sqrt{k!}}.$$

Evidently, $\|B_\lambda^m\| \leq N_{2p}^m(B_\lambda)$. Now relation (8.7) implies

$$\|B_\lambda^m\| \leq \frac{N_{2p}^m(V)}{\rho^m(A, \lambda)}.$$

Consequently,

$$\|B_\lambda^{pk+m}\| \leq \frac{N_{2p}^{kp+m}(V)}{\rho^{kp+m}(A, \lambda)\sqrt{k!}}.$$

Taking into account (8.4), we have

$$\|R_\lambda(A)\| \leq \|R_\lambda(D)\| \sum_{m=0}^{p-1} \sum_{k=0}^{j} \frac{N_{2p}^{kp+m}(V)}{\rho^{kp+m}(A, \lambda)\sqrt{k!}}.$$

Now (8.6) yields the required result. $\square$

Letting $j \to \infty$ in the last lemma, we easily get

Corollary 6.8.2 Let $A \in C_{2p}$ $(p = 1, 2, ...)$. Then

$$\|R_\lambda(A)\| \leq \sum_{m=0}^{p-1} \sum_{k=0}^{\infty} \frac{N_{2p}^{kp+m}(V)}{\rho^{pk+m+1}(A, \lambda)\sqrt{k!}} \quad (\lambda \notin \sigma(A)),$$

where V is the nilpotent part of A.

Proof of Theorem 6.7.1: Due to inequality (2.3) $N_{2p}(D) \leq N_{2p}(A)$. Thus, the triangular representation implies

$$N_{2p}(V) \leq N_{2p}(A) + N_{2p}(D) \leq 2N_{2p}(A). \tag{8.8}$$

Now, the required result follows from the previous lemma. $\square$

To prove Theorem 6.7.3, we need the gollowing

Lemma 6.8.3 *Under condition (7.1), let V be the nilpotent part of A. Then there are constants $a_0, b_0 > 0$, such that the inequality*

$$\|R_\lambda(A)\| \leq a_0 \sum_{m=0}^{p-1} \frac{(N_{2p}(V))^m}{\rho^{m+1}(A,\lambda)} \, exp \left[\frac{b_0(N_{2p}(V))^{2p}}{\rho^{2p}(A,\lambda)}\right] \ (\lambda \notin \sigma(A))$$

holds. These constants can be taken as in (4.8) or in (4.9).

Proof: According to (8.3),

$$\|(A - \lambda I)^{-1}\| \leq \|(D - \lambda I)^{-1}\| \sum_{k=0}^{p-1} \|B_\lambda\|^k \|(I - B_\lambda^p)^{-1}\|. \tag{8.9}$$

It is not hard to check that B_λ^p is a Volterra operator (see also Lemma 7.3.4) below. Moreover, $B_\lambda^p \in C_2$. Due to Theorem 6.4.2,

$$\|(I - B_\lambda^p)^{-1}\| \leq e^{1/2} e^{N_2^2(B_\lambda^p)/2}. \tag{8.10}$$

But

$$N_2(B_\lambda^p) \leq N_{2p}^p(B_\lambda) \leq N_{2p}^p(V)\|R_\lambda(D)\|^p.$$

In addition, (8.6) implies that $N_2(B_\lambda^p) \leq N_{2p}^p(A)\rho^{-p}(A,\lambda)$. Now relations (8.7), (8.9) and (8.10) yield the required result. $\square$

The assertion of Theorem 6.7.3 follows from the previous lemma and (8.8).

6.9 Regular Functions of Hilbert-Schmidt Operators

Let A be a bounded linear operator acting in a separable Hilbert space H and f be a scalar-valued function, which is analytic on a neighborhood of $\sigma(A)$. Let a contour C consist of a finite number of rectifiable Jordan curves, oriented in the positive sense customary in the theory of complex variables. Suppose that C is the boundary of an open set $M \supset \sigma(A)$ and $M \cup C$ is contained in the domain of analycity of f. We define $f(A)$ by the equality

$$f(A) = -\frac{1}{2\pi i} \int_C f(\lambda) R_\lambda(A) d\lambda \tag{9.1}$$

(see the book by Dunford and Schwartz (1966, p. 568)).

Theorem 6.9.1 *Let A be a Hilbert-Schmidt operator and let f be a holomorphic function on a neighborhood of the closed convex hull $co(A)$ of the spectrum of A. Then*

$$\|f(A)\| \leq \sum_{k=0}^{\infty} \sup_{\lambda \in co(A)} |f^{(k)}(\lambda)| \frac{g^k(A)}{(k!)^{3/2}}. \tag{9.2}$$

Proof: Thanks to Corollary 6.3.6, there is a sequence $\{A_n\}$ of operators having n-dimensional ranges, such that relations (4.5) hold. Corollary 2.7.2 implies

$$\|f(A_n)\| \leq \sum_{k=0}^{n-1} \sup_{\lambda \in co(A_n)} |f^{(k)}(\lambda)| \frac{g^k(A_n)}{(k!)^{3/2}}. \tag{9.3}$$

Due to the well-known Lemma VII.6.5 from (Dunford and Schwartz, 1966) we have

$$\|f(A_n) - f(A)\| \to 0 \text{ as } n \to \infty.$$

Letting $n \to \infty$ in (9.3), due to Lemma 6.3.6, we arrive at the stated result.
□

Theorem 6.9.1 is precise: inequality (9.2) becomes the equality

$$\|f(A)\| = \sup_{\mu \in \sigma(A)} |f(\mu)|,$$

if A is a normal operator and

$$\sup_{\lambda \in co(A)} |f(\lambda)| = \sup_{\lambda \in \sigma(A)} |f(\lambda)|,$$

because $g(A) = 0$ in this case.

Corollary 6.9.2 *Let A be a Hilbert-Schmidt operator. Then*

$$\|e^{At}\| \leq e^{\alpha(A)t} \sum_{k=0}^{\infty} \frac{t^k g^k(A)}{(k!)^{3/2}} \text{ for all } t \geq 0,$$

where $\alpha(A) = \sup Re\, \sigma(A)$. In addition,

$$\|A^m\| \leq \sum_{k=0}^{m} \frac{m! r_s^{m-k}(A) g^k(A)}{(m-k)!(k!)^{3/2}} \quad (m = 1, 2, ...).$$

Recall that $r_s(A)$ is the spectral radius of operator A. In particular, if $V \in C_2$ is a Volterra operator, then

$$\|V^m\| \leq \frac{N_2^m(V)}{\sqrt{m!}} \quad (m = 1, 2, ...). \tag{9.4}$$

Note that an independent proof of inequality (9.4) can be found in Section 2.3 of the book (Gil', 1995). In addition, that inequality allows us to estimate a power of a Volterra von Neumann - Schatten operator.

Lemma 6.9.3 *For some integer $p \geq 1$, let $V \in C_{2p}$ be a Volterra operator. Then*

$$\|V^{kp+m}\| \leq \frac{N_{2p}^{pk+m}(V)}{\sqrt{k!}} \quad (k = 0, 1, 2, ...; \, m = 0, ..., p-1). \tag{9.5}$$

Proof: Relation (2.4) implies $N_2(V^p) \leq N_{2p}^p(V)$. Thus $V^p \in C_2$. Due to (9.4)

$$\|V^{pk}\| \leq \frac{N_2^k(V^p)}{\sqrt{k!}} \leq \frac{N_{2p}^{kp}(V)}{\sqrt{k!}}.$$

Since $\|V^m\| \leq N_{2p}^m(V)$,

$$\|V^{pk+m}\| \leq \|V^m\| \frac{N_{2p}^{kp}(V)}{\sqrt{k!}} \leq \frac{N_{2p}^{kp+m}(V)}{\sqrt{k!}},$$

as claimed. $\square$

Inequality (9.5) can be rewritten in the following way:

Corollary 6.9.4 *For some integer $p \geq 1$, let $V \in C_{2p}$ be a Volterra operator. Then*

$$\|V^j\| \leq \theta_j^{(p)} N_{2p}^j(V) \quad (j = 1, 2, ...).$$

Recall that $\theta_j^{(p)}$ is given in Section 6.7.

6.10   A Relation between Determinants and Resolvents

Let $A \in C_2$. Then the generalized determinant

$$det_2(I - A) := \prod_{k=1}^{\infty} (1 - \lambda_k) e^{\lambda_k} \quad (\lambda_k \equiv \lambda_k(A))$$

is finite (Dunford and Schwartz, 1963, p. 1038). The following theorem is due to Carleman (see the next section), but we suggest a new proof and correct a misprint in Theorem XI.6.27 of the book (Dunford and Schwartz, 1963).

Theorem 6.10.1 *Let $A \in C_2$. Then*

$$\|(I - A)^{-1} det_2(I - A)\| \leq exp\,[(N_2^2(A) + 1)/2]. \tag{10.1}$$

Proof: Thanks to Theorem 2.11.1,

$$\|(I-A_n)^{-1} det\,(I-A_n)\| \leq [1 + \frac{1}{n-1}\,(N_2^2(A_n) - 2Re\,Trace\,(A_n) + 1)]^{(n-1)/2}$$

for any n-dimensional operator A_n. Hence,

$$\|(I - A_n)^{-1} det\,(I - A_n)\| \leq exp\,[(N_2^2(A_n) - 2Re\,Trace\,(A_n) + 1)/2].$$

Rewrite this relation as

$$\|(I - A_n)^{-1} det\,(I - A_n) exp\,[Re\,Trace\,(A_n)]\| \leq exp\,[(N_2^2(A_n) + 1)/2].$$

Or

$$\|(I - A_n)^{-1}\| \prod_{k=1}^{n} |(1 - \lambda_k(A_n))e^{\lambda_k(A_n)}| \leq exp\,[(N_2^2(A_n) + 1)/2].$$

Hence,

$$\|(I - A_n)^{-1}\,det_2(I - A_n)\| \leq exp\,[(N_2^2(A_n) + 1)/2]. \qquad (10.2)$$

Let A_n, $n = 1, 2, ...$ converge to A in the norm $N_2(.)$ and satisfy conditions (3.4). Then

$$det_2(I - A_n) \to det_2(I - A).$$

This finishes the proof. $\square$
Replacing in (10.1) A by $\lambda^{-1}A$, we get

Corollary 6.10.2 *Let $A \in C_2$. Then*

$$\|(\lambda I - A)^{-1}\,det_2(I - \lambda^{-1}A)\| \leq \frac{1}{|\lambda|}\,exp\,[\frac{1}{2} + \frac{N_2^2(A)}{2|\lambda|^2}]\,\,(\lambda \notin \sigma(A)). \quad (10.3)$$

In particular, if $V \in C_2$ is quasinilpotent, then

$$\|(\lambda I - V)^{-1}\| \leq \frac{1}{|\lambda|}\,exp\,[\frac{1}{2} + \frac{N_2^2(V)}{2|\lambda|^2}]\,\,(\lambda \neq 0). \qquad (10.4)$$

Moreover, relation (10.4) implies.

Corollary 6.10.3 *Let $A \in C_2$. Then*

$$\|R_\lambda(A)\| \leq \frac{1}{\rho(A, \lambda)}exp\,[\frac{1}{2} + \frac{g^2(A)}{2\rho^2(A, \lambda)}]\,\,(\lambda \notin \sigma(A)). \qquad (10.5)$$

Indeed, due to (3.2), $R_\lambda(A) = R_\lambda(D)(I + VR_\lambda(D))^{-1}$. Lemma 7.3.4 below yields that $VR_\lambda(D)$ is a Volterra operator. So according to (10.4),

$$\|(I + VR_\lambda(D))^{-1}\|\|(I + VR_\lambda(D))^{-1}\| \leq exp\,[\frac{1}{2} + \frac{N_2^2(VR_\lambda(D))}{2}] \leq$$

$$exp\,[\frac{1}{2} + \frac{N_2^2(V)\|R_\lambda(D)\|^2}{2}].$$

But due to Lemma 6.5.2, $N_2(V) = g(A)$. Thus,

$$\|R_\lambda(A)\| \leq \|R_\lambda(D)(I + VR_\lambda(D))^{-1}\| \leq \frac{1}{\rho(A, \lambda)}exp\,[\frac{1}{2} + \frac{g^2(A)}{2\rho^2(A, \lambda)}],$$

as claimed.
Note that Corollary 6.10.3 gives us an additional proof of Theorem 6.4.2.

6.11   Notes

Theorems 6.4.1, 6.7.1 and 6.9.1 were derived in the papers (Gil', 1979a), (Gil', 1992) and (Gil', 1979b), respectively (see also (Gil', 1995, Chapter 2)), but in the present chapter we suggest the new proofs. Theorems 6.4.2 and 6.7.3 are probably new.

In the book (Dunford and Schwartz, 1963, p. 1038), instead of (10.3), it is erroneously stated that

$$\|(\lambda I - A)^{-1} \, det_2(I - \lambda^{-1}A)\| \leq |\lambda| \, exp \, [\frac{1}{2} + \frac{N_2^2(A)}{2|\lambda|^2}].$$

Note that the very interesting estimates for the resolvents of operators from C_p are established in the papers (Dechevski and Persson, 1994 and 1996).

References

[1] Ahiezer, N. I. and Glazman, I. M. (1981). *Theory of Linear Operators in a Hilbert Space*. Pitman Advanced Publishing Program, Boston.

[2] Brodskii, M. S. (1971). *Triangular and Jordan Representations of Linear Operators*, Transl. Math. Mongr., v. 32, Amer. Math. Soc., Providence, R. I.

[3] Dechevski, L. T. and Persson, L. E. (1994). Sharp generalized Carleman inequalities with minimal information about the spectrum, *Math. Nachr.*, **168**, 61-77.

[4] Dechevski, L. T. and Persson, L. E. (1996). On sharpness, applications and generalizations of some Carleman type inequalities, *Tôhuku Math. J.*, **48**, 1-22.

[5] Dunford, N. and Schwartz, J. T. (1966). *Linear Operators, part I. General Theory*. Interscience publishers, New York.

[6] Dunford, N. and Schwartz, J. T. (1963). *Linear Operators, part II. Spectral Theory*. Interscience publishers, New York, London.

[7] Gil', M. I. (1979a). An estimate for norms of resolvent of completely continuous operators, *Mathematical Notes*, **26** , 849-851.

[8] Gil', M. I. (1979b). Estimates for norms of functions of a Hilbert-Schmidt operator (in Russian), *Izvestiya VUZ, Matematika*, **23**, 14-19. English translation in *Soviet Math.*, **23**, 13-19.

[9] Gil' , M. I. (1992). On estimate for the norm of a function of a quasi-hermitian operator, *Studia Mathematica*, **103(1)**, 17-24.

[10] Gil', M. I. (1995). *Norm Estimations for Operator-valued Functions and Applications.* Marcel Dekker, Inc., New York.

[11] Gohberg, I. C. and Krein, M. G. (1969). *Introduction to the Theory of Linear Nonselfadjoint Operators,* Trans. Mathem. Monographs, v. 18, Amer. Math. Soc., Providence, R. I.

[12] Gohberg, I. C. and Krein, M. G. (1970) . *Theory and Applications of Volterra Operators in Hilbert Space,* Trans. Mathem. Monographs, v. 24, Amer. Math. Soc., Providence, R. I.

[13] Pietsch, A. (1988). *Eigenvalues and s-numbers,* Cambridge University Press, Cambridge.

7. Functions of Non-compact Operators

The present chapter is concerned with the estimates for the norms of resolvents and analytic functions of so called P-triangular operators. Roughly speaking, a P-triangular operator is a sum of a normal operator and a compact quasinilpotent one, having a sufficiently rich set of invariant subspaces. In particular, we consider the following classes of P-triangular operators: operators whose Hermitian components are compact operators, and operators, which are represented as sums of unitary operators and compact ones.

7.1 Terminology

Let A be a linear operator acting in a separable Hilbert space H. Let there be a linear manifold $Dom\ (A)$, such that the relation $f \in Dom\ (A)$ implies $Af \in H$. Then the set $Dom\ (A)$ is called *the domain of* A. Let $Dom\ (A)$ be dense in H. Then the set of vectors g satisfying

$$|(Af, g)| \leq c\|f\| \text{ for all } f \in Dom\ (A)$$

with a constant c is the domain of *the adjoint operator* A^* and, besides,

$$(Af, g) = (f, A^*g) \text{ for all } f \in Dom\ (A) \text{ and } g \in Dom\ (A^*).$$

An unbounded operator A is selfadjoint, if

$$Dom\ (A) = Dom\ (A^*) \text{ and } Ah = A^*h \ \ (h \in Dom\ (A)).$$

An unbounded selfadjoint operator possesses an unbounded real spectrum. An unbounded operator A is normal, if

$$Dom\ (AA^*) = Dom\ (A^*A) \text{ and } AA^*h = A^*Ah \ \ (h \in Dom\ (A^*A)).$$

An unbounded normal operator has an unbounded spectrum.

Definition 7.1.1 *Let A be a linear operator in H. Then A is said to be a quasi-normal operator, if it is a sum of a normal operator and a compact one.*

Operator A is said to be a quasi-Hermitian operator, if it is a sum of a selfadjoint operator and a compact one.

Let A be bounded and $A_I \equiv (A - A^*)/2i \in C_p$ $(p \geq 1)$. Then, as it is well-known (see e.g. (Gohberg and Krein, 1969, Section II.6)), the nonreal spectrum consists of the eigenvalues having finite multiplicities, and

$$\sum_{k=1}^{\infty} |Im\ \lambda_j(A)|^p \leq \sum_{k=1}^{\infty} |\lambda_j(A_I)|^p \quad (n = 1, 2, \ldots) \tag{1.1}$$

where $\lambda_j(A)$ and $\lambda_j(A_I)$ are the eigenvalues with their multiplicities of A and A_I, respectively.

7.2 P-Triangular Operators

A family of orthogonal projectors $P(t)$ in H (i.e. $P^2(t) = P(t)$ and $P^*(t) = P(t)$) defined on a (finite or infinite) segment $[a, b]$ of the real axis is a *an orthogonal resolution of the identity* if for all $t, s \in [a, b]$,

$$P(a) = 0,\ P(b) = I \equiv \text{ the unit operator and } P(t)P(s) = P(min(t, s))$$

An orthogonal resolution of the identity $P(t)$ is left-continuous, if $P(t - 0) = P(t)$ for all $t \in (a, b]$ in the sense of the strong topology.

Definition 7.2.1 *Let $P(t)$ be a left-continuous orthogonal resolution of the identity in H defined on a (finite or infinite) real segment $[a, b]$. Then $P(.)$ will be called a maximal resolution of the identity (m.r.i.), if its every gap $P(t_0 + 0) - P(t_0)$ (if it exists) is one-dimensional.*

Moreover, we will say that an m.r.i. $P(.)$ belongs to a linear operator A (or A has an m.r.i. $P(.)$), if

$$P(t)AP(t)h = AP(t)h \text{ for all } t \in [a, b] \text{ and } h \in Dom\ (A). \tag{2.1}$$

Recall that a linear operator V is called a Volterra one, if it is compact and quasinilpotent.

Definition 7.2.2 *Let a linear generally unbounded operator A have an m.r.i. $P(.)$ defined on $[a, b]$. In addition, let*

$$A = D + V, \tag{2.2}$$

where D is a normal operator and V is a Volterra one, having the following properties:

$$P(t)VP(t) = VP(t) \quad (t \in [a, b]) \tag{2.3}$$

and

$$DP(t)h = P(t)Dh \quad (t \in [a, b], \ h \in Dom \ (A)). \tag{2.4}$$

Then A will be called a P-triangular operator. In addition, equality (2.2), and operators D and V will be called the triangular representation, diagonal part and nilpotent part of A, respectively.

Clearly, this definition is in accordance with the definition of the triangular representation of compact operators (see Section 6.3).

7.3 Some Properties of Volterra Operators

Lemma 7.3.1 *Let a compact operator V in H, have a maximal orthogonal resolution of the identity $P(t)$ ($a \leq t \leq b$) (that is, condition (2.3) hold). If, in addition,*

$$(P(t_0 + 0) - P(t_0))V(P(t_0 + 0) - P(t_0)) = 0 \tag{3.1}$$

for every gap $P(t_0 + 0) - P(t_0)$ of $P(t)$ (if it exists), then V is a Volterra operator.

Proof: Since the set of the values of $P(.)$ is a maximal chain of projectors, the required result is due to Corollary 1 to Theorem 17.1 of the book by Brodskii (1971). □

In particular, if $P(t)$ is continuous in t in the strong topology and (2.3) holds, then V is a Volterra operator. We also need the following

Lemma 7.3.2 *Let V be a Volterra operator in H, and $P(t)$ a maximal orthogonal resolution of the identity satisfying equality (2.3). Then equality (3.1) holds for every gap $P(t_0 + 0) - P(t_0)$ of $P(t)$ (if it exists).*

Proof: Since the set of the values of $P(.)$ is a maximal chain of projectors, the required result is due to the well-known equality (I.3.1) from the book by Gohberg and Krein (1970). □

Lemma 7.3.3 *Let V and B be bounded linear operators in H having the same m.r.i. $P(.)$. In addition, let V be a Volterra operator. Then VB and BV are Volterra operators, and $P(.)$ is their m.r.i.*

Proof: It is obvious that

$$P(t)VBP(t) = VP(t)BP(t) = VBP(t). \tag{3.2}$$

Now let $Q = P(t_0 + 0) - P(t_0)$ be a gap of $P(t)$. Then according to Lemma 7.3.2 equality (3.1) holds. Further, we have

$$QVBQ = QVB(P(t_0 + 0) - P(t_0)) =$$

$$QV[P(t_0 + 0)BP(t_0 + 0) - P(t_0)BP(t_0)] =$$

$$QV[(P(t_0) + Q)B(P(t_0) + Q) - P(t_0)BP(t_0)] =$$

$$QV[P(t_0)BQ + QBP(t_0)].$$

Since $QP(t_0) = 0$ and $P(t)$ projects onto the invariant subspaces, we obtain $QVBQ = 0$. Due to Lemma 7.3.1 this relation and equality (3.2) imply that VB is a Volterra operator. Similarly we can prove that BV is a Volterra one. $\square$

Lemma 7.3.4 *Let A be a (generally unbounded) P-triangular operator. Let V and D be the nilpotent and diagonal parts of A, respectively. Then for any regular point λ of D, the operators $VR_\lambda(D)$ and $R_\lambda(D)V$ are Volterra ones. Besides, A, $VR_\lambda(D)$ and $R_\lambda(D)V$ have the same m.r.i.*

Proof: Due to (2.4)

$$P(t)R_\lambda(D) = R_\lambda(D)P(t) \text{ for all } t \in [a, b].$$

Now Lemma 7.3.3 ensures the required result. $\square$

7.4   Powers of Volterra Operators

Let Y be a *a norm ideal of compact linear operators in H*. That is, Y is algebraically a two- sided ideal, which is complete in an auxiliary norm $|\cdot|_Y$ for which $|CB|_Y$ and $|BC|_Y$ are both dominated by $\|C\| |B|_Y$.

In the sequel we suppose that there are positive numbers θ_k $(k \in \mathbf{N})$, with

$$\theta_k^{1/k} \to 0 \text{ as } k \to \infty,$$

such that

$$\|V^k\| \leq \theta_k |V|_Y^k \tag{4.1}$$

for an arbitrary Volterra operator

$$V \in Y. \tag{4.2}$$

Recall that C_{2p} $(p = 1, 2, ...)$ is the von Neumann-Schatten ideal of compact operators with the finite ideal norm

$$N_{2p}(K) \equiv [Trace \ (K^*K)^p]^{1/2p} \ (K \in C_{2p}).$$

Let $V \in C_{2p}$ be a Volterra operator. Then due to Corollary 6.9.4,

$$\|V^j\| \le \theta_j^{(p)} N_{2p}^j(V) \quad (j = 1, 2, ...) \tag{4.3}$$

where

$$\theta_j^{(p)} = \frac{1}{\sqrt{[j/p]!}}$$

and $[x]$ means the integer part of a positive number x. Inequality (4.3) can be written as

$$\|V^{kp+m}\| \le \frac{N_{2p}^{pk+m}(V)}{\sqrt{k!}} \quad (k = 0, 1, 2, ...; \ m = 0, ..., p - 1). \tag{4.4}$$

7.5 Resolvents of P-Triangular Operators

Lemma 7.5.1 *Let A be a P-triangular operator. Then $\sigma(A) = \sigma(D)$, where D is the diagonal part of A.*

Proof: Let λ be a regular point of the operator D. According to the triangular representation (2.2) we obtain

$$R_\lambda(A) = (D + V - \lambda I)^{-1} = R_\lambda(D)(I + V R_\lambda(D))^{-1}. \tag{5.1}$$

Operator $V R_\lambda(D)$ for a regular point λ of the operator D is a Volterra one due to Lemma 7.3.4. Therefore,

$$(I + V R_\lambda(D))^{-1} = \sum_{k=0}^\infty (V R_\lambda(D))^k (-1)^k$$

and the series converges in the operator norm. Thus,

$$R_\lambda(A) = R_\lambda(D) \sum_{k=0}^\infty (V R_\lambda(D))^k (-1)^k. \tag{5.2}$$

Hence, it follows that λ is the regular point of A.

Conversely let $\lambda \notin \sigma(A)$. According to the triangular representation (2.2) we obtain

$$R_\lambda(D) = (A - V - \lambda I)^{-1} = R_\lambda(A)(I - V R_\lambda(A))^{-1}.$$

Operator $V R_\lambda(A)$ for a regular point λ of operator A is a Volterra one due to Lemma 7.3.3. So

$$(I - V R_\lambda(A))^{-1} = \sum_{k=0}^\infty (V R_\lambda(A))^k$$

and the series converges in the operator norm. Thus,

$$R_\lambda(D) = R_\lambda(A) \sum_{k=0}^{\infty} (V R_\lambda(A))^k.$$

Hence, it follows that λ is the regular point of D. This finishes the proof. $\square$

Furthermore, for a natural number m and a $z \in (0, \infty)$, under (4.2), put

$$J_Y(V, m, z) = \sum_{k=0}^{m-1} \frac{\theta_k |V|_Y^k}{z^{k+1}}. \tag{5.3}$$

Definition 7.5.2 *A number $ni(V)$ is called the nilpotency index of a nilpotent operator V, if $V^{ni(V)} = 0 \neq V^{ni(V)-1}$. If V is quasinilpotent but not nilpotent we write $ni(V) = \infty$.*

Everywhere below one can replace $ni(V)$ by ∞.

Theorem 7.5.3 *Let A be a P-triangular operator and let its nilpotent part V belong to a norm ideal Y with the property (4.1). Then*

$$\|R_\lambda(A)\| \leq J_Y(V, \nu(\lambda), \rho(A, \lambda)) := \sum_{k=0}^{\nu(\lambda)-1} \frac{\theta_k |V|_Y^k}{\rho^{k+1}(A, \lambda)} \tag{5.4}$$

for all regular λ of A. Here $\nu(\lambda) = ni(V R_\lambda(D))$, and D is the diagonal part of A.

Proof: Due to Lemma 7.3.4, $V R_\lambda(D) \in Y$ is a Volterra operator. So according to (4.1),

$$\|(V R_\lambda(D))^k\| \leq \theta_k |V R_\lambda(D)|_Y^k.$$

But

$$|V R_\lambda(D)|_Y \leq |V|_Y \|R_\lambda(D)\|,$$

and thanks to Lemma 7.5.1,

$$\|R_\lambda(D)\| = \frac{1}{\rho(D, \lambda)} = \frac{1}{\rho(A, \lambda)}.$$

So

$$\|(V R_\lambda(D))^k\| \leq \frac{\theta_k |V|_Y^k}{\rho^k(A, \lambda)}.$$

Relation (5.2) implies

$$\|R_\lambda(A)\| \leq \|R_\lambda(D)\| \sum_{k=0}^{\nu(\lambda)-1} \|(V R_\lambda(D))^k\| \leq$$

$$\frac{1}{\rho(A,\lambda)} \sum_{k=0}^{\nu(\lambda)-1} \frac{\theta_k |V|_Y^k}{\rho^k(A,\lambda)},$$

as claimed. $\square$

For a natural number m, a Volterra operator $V \in C_{2p}$ and a $z \in (0,\infty)$, put

$$\tilde{J}_p(V,m,z) := \sum_{k=0}^{m-1} \frac{\theta_k^{(p)} N_{2p}^k(V)}{z^{k+1}}. \tag{5.5}$$

Theorem 7.5.3 and inequality (4.3) yield

Corollary 7.5.4 *Let A be a P-triangular operator and its nilpotent part $V \in C_{2p}$ for an integer $p \geq 1$. Then*

$$\|R_\lambda(A)\| \leq \tilde{J}_p(V,\nu(\lambda),\rho(A,\lambda)) \equiv \sum_{k=0}^{\nu(\lambda)-1} \frac{\theta_k^{(p)} N_{2p}^k(V)}{\rho^{k+1}(A,\lambda)} \quad (\lambda \notin \sigma(A)). \tag{5.6}$$

In particular, let A be a P-triangular operator, whose nilpotent part V is a Hilbert-Schmidt operator. Then due to the previous corollary

$$\|R_\lambda(A)\| \leq \tilde{J}_1(V,\nu(\lambda),\rho(A,\lambda)) \equiv \sum_{k=0}^{\nu(\lambda)-1} \frac{N_2^k(V)}{\rho^{k+1}(A,\lambda)\sqrt{k!}} \quad (\lambda \notin \sigma(A)). \tag{5.7}$$

Furthermore, inequality (5.6) implies

$$\|R_\lambda(A)\| \leq \sum_{j=0}^{p-1} \sum_{k=0}^{\infty} \frac{N_{2p}^{pk+j}(V)}{\rho^{pk+j+1}(A,\lambda)\sqrt{k!}}. \tag{5.8}$$

Theorem 7.5.5 *Let A be a P-triangular operator and its nilpotent part $V \in C_{2p}$ for some integer $p \geq 1$. Then there are constants $a_0, b_0 > 0$, such that*

$$\|R_\lambda(A)\| \leq a_0 \sum_{j=0}^{p-1} \frac{N_{2p}^j(V)}{\rho^{j+1}(A,\lambda)} \, exp\, [\frac{b_0 \, N_{2p}^{2p}(V)}{\rho^{2p}(A,\lambda)}] \quad (\lambda \notin \sigma(A)). \tag{5.9}$$

One can take

$$a_0 = \sqrt{\frac{c}{c-1}} \;\; and \;\; b_0 = \frac{c}{2} \;\; for \; any \; c > 1; \; in \; particular, \; a_0 = \sqrt{2} \; and \; b_0 = 1, \tag{5.10}$$

or

$$a_0 = e^{1/2} \;\; and \;\; b_0 = 1/2. \tag{5.11}$$

Proof: Due to Lemma 6.8.3, for any Volterra operator $V \in C_{2p}$,

$$\|(I - V)^{-1}\| \leq a_0 \sum_{j=0}^{p-1} N_{2p}^j(V) \, exp \, [\, b_0 N_{2p}^{2p}(V)].$$

But $V R_\lambda(D)$ is a Volterra operator due to Lemma 7.3.4. Hence,

$$\|(I + V R_\lambda(D))^{-1}\| \leq a_0 \sum_{j=0}^{p-1} N_{2p}^j(V R_\lambda(D)) \, exp \, [\, b_0 N_{2p}^{2p}(V R_\lambda(D))] \leq$$

$$\leq a_0 \sum_{j=0}^{p-1} \frac{N_{2p}^j(V)}{\rho^j(A, \lambda)} exp \, [\, b_0 \frac{N_{2p}^{2p}(V)}{\rho^{2p}(A, \lambda)}].$$

Now (5.1) implies the required result. □

7.6　Triangular Representations of Quasi-Hermitian Operators

Theorem 7.6.1 *Let a linear generally unbounded operator A satisfy the conditions Dom (A^*) = Dom (A) and*

$$A_I = (A - A^*)/2i \in C_p \quad (1 < p < \infty). \tag{6.1}$$

Then A admits the triangular representation (2.2).

Proof: First, let A be bounded. Then, as it is proved by L. de Branges (1965a, p. 69), under condition (6.1), there are a maximal orthogonal resolution of the identity $P(t)$ defined on a finite real segment $[a, b]$ and a real nondecreasing function $h(t)$, such that

$$A = \int_a^b h(t) dP(t) + 2i \int_a^b P(t) A_I dP(t). \tag{6.2}$$

The second integral in (6.2) is understood as the limit in the operator norm of the operator Stieltjes sums

$$L_n = \frac{1}{2} \sum_{k=1}^n [P(t_{k-1}) + P(t_k)] A_I \Delta P_k,$$

where

$$t_k = t_k^{(n)}; \ \Delta P_k = P(t_k) - P(t_{k-1}); \ a = t_0 < t_1 \ldots < t_n = b.$$

We can write $L_n = W_n + T_n$ with

$$W_n = \sum_{k=1}^{n} P(t_{k-1}) A_I \Delta P_k \text{ and } T_n = \frac{1}{2} \sum_{k=1}^{n} \Delta P_k A_I \Delta P_k. \qquad (6.3)$$

The sequence $\{T_n\}$ converges in the operator norm due to the well-known Lemma I.5.1 from the book (Gohberg and Krein, 1970). We denote its limit by T. Clearly, T is selfadjoint and $P(t)T = TP(t)$ for all $t \in [a,b]$. Put

$$D = \int_a^b h(t)dP(t) + 2iT. \qquad (6.4)$$

Then D is normal and satisfies condition (2.4).

Furthermore, it can be directly checked that W_n is a nilpotent operator: $(W_n)^n = 0$. Besides, the sequence $\{W_n\}$ converges in the operator norm, because the second integral in (6.2) and $\{T_n\}$ converge in this norm. We denote the limit of the sequence $\{2iW_n\}$ by V. It is a Volterra operator, since the limit in the operator norm of a sequence of Volterra operators is a Volterra one (see for instance Lemma 2.17.1 from the book (Brodskii, 1971)). From this we easily obtain relations (2.1)-(2.4). So, for bounded operators the theorem is proved.

Now let A be unbounded. Due to De Branges (1965a, p. 69), there is a maximal orthogonal resolution of the identity $P(t)$, $-\infty \le t \le \infty$ and nondecreasing functions $h(t)$, such that under (6.1), the representation

$$A = \int_{-\infty}^{\infty} h(t)dP(t) + 2i \int_{-\infty}^{\infty} P(t)A_I dP(t) \qquad (6.5)$$

holds. The integrals in (6.5) are understood as the limits of corresponding integrals from (6.2) when $a \to -\infty$ and $b \to \infty$. Besides, the first integral in (6.5) is the strong limit on $Dom(A)$ of the first integral in (6.2), and the second integral in (6.5) is the limit in the uniform operator topology of the second integral in (6.2). Take

$$A_n = A(P(n) - P(-n)). \qquad (6.6)$$

Clearly, A_n is bounded and has the property (6.1). So as it was proved above, according to (6.5) it has a triangular representation with the m.r.i. $P(.)$. Hence, letting $n \to \infty$, we get the required result. $\square$

Corollary 7.6.2 *Let A be an unbounded operator with the property (6.1). Then it is P-triangular and there is a sequence of bounded P-triangular operators A_n, such that $A_n - A_n^* \in C_p$,*

$$\sigma(A_n) \subseteq \sigma(A), \ N_p(A_n - A^*) \to 0 \text{ and } N_p(V_n - V) \to 0$$

as $n \to 0$, where V_n and V are the nilpotent part of A_n and A, respectively. Moreover, operators A and A_n ($n = 1, 2, ...$) have the same m.r.i.

Indeed, taking A_n as in (6.6), we arrive at the result due to (2.2) and the previous theorem.

7.7 Resolvents of Operators with Hilbert-Schmidt Hermitian Components

In this section we obtain an estimate for the norm of the resolvent of a generally unbounded quasi-Hermitian operator under the conditions $Dom\,(A) = Dom\,(A^*)$ and

$$A_I = (A - A^*)/2i \text{ is a Hilbert-Schmidt operator.} \tag{7.1}$$

Let us introduce the quantity

$$g_I(A) \equiv \sqrt{2}\,[N_2^2(A_I) - \sum_{k=1}^{\infty}(Im\,\lambda_k(A))^2]^{1/2}. \tag{7.2}$$

Theorem 7.7.1 *Let condition (7.1) hold. Then*

$$\|R_\lambda(A)\| \leq \sum_{k=0}^{\infty} \frac{g_I^k(A)}{\rho^{k+1}(A,\lambda)\sqrt{k!}} \quad (\lambda \notin \sigma(A)). \tag{7.3}$$

Moreover, there are constants $a_0, b_0 > 0$, such that

$$\|R_\lambda(A)\| \leq \frac{a_0}{\rho(A,\lambda)}exp\,[\,\frac{b_0\,g_I^2(A)}{\rho^2(A,\lambda)}] \quad (\lambda \notin \sigma(A)). \tag{7.4}$$

These constants can be taken from (5.10) or from (5.11).

Here $\rho(A,\lambda)$ is the distance between the spectrum $\sigma(A)$ of A and a complex point λ, again.

To prove this theorem we need the following

Lemma 7.7.2 *Let an operator A satisfy the condition (7.1). Then it admits the triangular representation (due to Theorem 7.6.1). Moreover, $N_2(V) = g_I(A)$, where V is the nilpotent part of A.*

Proof: First, assume that A is a bounded operator. Let D be the diagonal part of A. From the triangular representation (2.2) it follows that

$$-4Tr\,A_I^2 = Tr\,(A - A^*)^2 = Tr\,(D + V - D^* - V^*)^2.$$

By Theorem 6.2.2 and Lemma 7.3.3, $Trace\,VD^* = Trace\,V^*D = 0$. Hence, omitting simple calculations, we obtain

$$-4Tr\,A_I^2 = Tr\,(D - D^*)^2 + Tr\,(V - V^*)^2.$$

That is, $N_2^2(A_I) = N_2^2(V_I) + N_2^2(D_I)$, where $V_I = (V - V^*)/2i$ and $D_I = (D - D^*)/2i$. Taking into account Lemma 6.5.1, we arrive at the equality

$$2N^2(A_I) - 2N^2(D_I) = N^2(V).$$

Recall that the nonreal spectrum of a quasi-Hermitian operator consists of isolated eigenvalues. Besides, due to Lemma 7.5.1 $\sigma(A) = \sigma(D)$. Thus,

$$N^2(D_I) = \sum_{k=1}^{\infty} |Im\ \lambda_k(A)|^2,$$

and we arrive at the result, if A is bounded. If A is unbounded, then the required assertion is due to Corollary 7.6.2 and the just proved result for bounded operators. $\square$

Proof of Theorem 7.7.1: Inequality (7.3) follows from Corollary 7.5.4 and Lemma 7.7.2. Inequality (7.4) follows from Theorem 7.5.5 and Lemma 7.7.2. $\square$

7.8 Operators with the Property $A^p - (A^*)^p \in C_2$

Theorem 7.8.1 *Let a bounded linear operator A satisfy the condition*

$$A^p - (A^*)^p \in C_2 \quad (p = 2, 3, ...). \tag{8.1}$$

Then

$$\|R_\lambda(A)\| \le \|T_{\lambda,p}\| \sum_{k=0}^{\infty} \frac{g_I^k(A^p)}{\rho^{k+1}(A^p, \lambda^p)\sqrt{k!}} \quad (\lambda^p \notin \sigma(A^p)).$$

Here $\rho(A^p, \lambda^p)$ is the distance between the spectrum $\sigma(A^p)$ of A^p and λ^p, and

$$T_{\lambda,p} = \sum_{k=0}^{p-1} A^k \lambda^{p-k-1}.$$

Indeed, this result follows from Theorem 7.7.1 and the obvious relation

$$R_\lambda(A) = T_{\lambda,p}\ R_{\lambda^p}(A^p). \tag{8.2}$$

Moreover, relations (7.4) and (8.2) yield

Corollary 7.8.2 *Let condition (8.1) hold. Then there are constants $a_0, b_0 > 0$, such that*

$$\|R_\lambda(A)\| \le \frac{a_0\|T_{\lambda,p}\|}{\rho(A^p, \lambda^p)}\ exp\ [\frac{b_0\ g_I^2(A^p)}{\rho^2(A^p, \lambda^p)}] \quad (\lambda^p \notin \sigma(A^p)).$$

These constants can be taken from (5.10) or from (5.11).

7.9　Resolvents of Operators with Neumann - Schatten Hermitian Components

In this section we obtain estimates for the resolvent of an operator A, assuming that $Dom\,(A) = Dom\,(A^*)$ and

$$A_I := (A - A^*)/2i \in C_{2p} \text{ for some integer } p > 1. \qquad (9.1)$$

That is,

$$N_{2p}(A_I) = \sqrt[2p]{\sum_{j=1}^{\infty} \lambda_j^{2p}(A_I)} < \infty.$$

Put

$$\tilde{\beta}_p := \begin{cases} 2(1 + ctg\,(\frac{\pi}{4p})\,) & \text{if } p = 2^{m-1},\ m = 1, 2, \dots \\ 2(1 + \frac{2p}{exp(2/3)ln2}) & \text{otherwise} \end{cases} . \qquad (9.2)$$

Theorem 7.9.1 *Let condition (9.1) hold. Then there is a constant β_p, depending on p, only, such that*

$$\|R_\lambda(A)\| \le \sum_{m=0}^{p-1} \sum_{k=0}^{\infty} \frac{(\beta_p N_{2p}(A_I))^{kp+m}}{\rho^{pk+m+1}(A, \lambda)\sqrt{k!}} \quad (\lambda \notin \sigma(A)). \qquad (9.3)$$

Besides,

$$\beta_p \le \tilde{\beta}_p. \qquad (9.4)$$

Moreover, there are constants $a_0, b_0 > 0$, such that

$$\|R_\lambda(A)\| \le a_0 \sum_{m=0}^{p-1} \frac{(\beta_p N_{2p}(A_I))^m}{\rho^{m+1}(A, \lambda)} exp\,\Big[\frac{b_0(\beta_p N_{2p}(A_I))^{2p}}{\rho^{2p}(A, \lambda)}\Big] \quad (\lambda \notin \sigma(A)). \qquad (9.5)$$

Constants a_0, b_0 can be taken from (5.10) or from (5.11).

In order to prove this theorem we need the following result.

Lemma 7.9.2 *Let A satisfy condition (9.1). Then it admits the triangular representation (due to Theorem 7.6.1), and the nilpotent part V of A satisfies the relation*

$$N_{2p}(V) \le \beta_p N_{2p}(A_I).$$

Proof: Let $V \in C_p$ $(2 \le p < \infty)$ be a Volterra operator. Then due to the well-known Theorem III.6.2 from the book by Gohberg and Krein (1970, p. 118), there is a constant γ_p, depending on p, only, such that

$$N_p(V_I) \le \gamma_p N_p(V_R) \quad (V_I = (V + V^*)/2i,\ V_R := (V + V^*)/2). \qquad (9.6)$$

Besides, as it is proved in (Gohberg and Krein, 1970, pages 123 and 124),

$$\frac{p}{2\pi} \leq \gamma_p \leq \frac{p}{exp(2/3)ln2}.$$

Moreover,

$$\gamma_p = ctg \frac{\pi}{2p} \text{ if } p = 2^n \ (n = 1, 2, ...),$$

cf. (Gohberg and Krein, 1970, page 124). Take

$$\beta_p = 2(1 + \gamma_{2p}).$$

Now let D be the diagonal part of A. Let V_I , D_I be the imaginary components of V and D, respectively. According to (2.2) $A_I = V_I + D_I$. First, assume that A is a bounded operator. Due to (1.1), the condition $A_I \in C_{2p}$ entails the inequality $N_{2p}(D_I) \leq N_{2p}(A_I)$. Therefore,

$$N_{2p}(V_I) \leq N_{2p}(A_I) + N_{2p}(D_I) \leq 2N_{2p}(A_I).$$

Hence, due to (9.6),

$$N_{2p}(V) \leq N_{2p}(V_R) + N_{2p}(V_I) \leq N_{2p}(V_I)(1 + \gamma_{2p}) = \beta_p N_{2p}(A_I),$$

as claimed.

Now let A be unbounded. To obtain the stated result in this case it is sufficient to apply Corollary 7.6.2 and the just obtained result for bounded operators. $\square$

The assertion of Theorem 7.9.1 follows from (5.8), Theorem 7.5.5 and Lemma 7.9.2.

7.10   Regular Functions of Bounded Quasi-Hermitian Operators

Let A be a bounded linear operator in a H and let $f(z)$ be a scalar-valued function which is analytic on some neighborhood of $\sigma(A)$. Again, put

$$f(A) = -\frac{1}{2\pi i} \int_C f(\lambda) R_\lambda(A) d\lambda, \tag{10.1}$$

where C is a closed Jordan contour, surrounding $\sigma(A)$ and having the positive orientation with respect to $\sigma(A)$. Below we also consider unbounded operators.

Theorem 7.10.1 *Let a bounded linear operator A satisfy the condition*

$$A_I \text{ is a Hilbert-Schmidt operator.} \tag{10.2}$$

In addition, let f be a holomorphic function on a neighborhood of the closed convex hull $co(A)$ of $\sigma(A)$. Then

$$\|f(A)\| \leq \sum_{k=0}^{\infty} \sup_{\lambda \in co(A)} |f^{(k)}(\lambda)| \frac{g_I^k(A)}{(k!)^{3/2}}. \tag{10.3}$$

Recall that the quantity $g_I(A)$ is defined by the equality (7.2).

Theorem 7.10.1 is precise: the inequality (10.3) becomes the equality

$$\|f(A)\| = \sup_{\mu \in \sigma(A)} |f(\mu)|, \tag{10.4}$$

if A is a normal operator and

$$\sup_{\lambda \in co(A)} |f(\lambda)| = \sup_{\lambda \in \sigma(A)} |f(\lambda)|, \tag{10.5}$$

because $g_I(A) = 0$ for a normal operator A.

Example 7.10.2 *Let a bounded operator A satisfy condition (10.2). Then Theorem 7.10.1 gives us the inequality*

$$\|A^m\| \leq \sum_{k=0}^{m} \frac{m! r_s^{m-k}(A) g_I^k(A)}{(m-k)!(k!)^{3/2}}$$

for any integer $m \geq 1$. Recall that $r_s(A)$ is the spectral radius of A.

Example 7.10.3 *Let a bounded operator A satisfy the condition (10.2). Then Theorem 7.10.1 gives us the inequality*

$$\|e^{At}\| \leq e^{\alpha(A)t} \sum_{k=0}^{\infty} \frac{t^k g_I^k(A)}{(k!)^{3/2}} \text{ for all } t \geq 0, \tag{10.6}$$

where $\alpha(A) = \sup Re\, \sigma(A)$.

7.11   Proof of Theorem 7.10.1

Let P_k $(k = 0, ..., n)$ be a finite chain of orthogonal projectors:

$$0 = Range(P_0) \subset Range(P_1) \subset .... \subset Range(P_n) = H.$$

We need the following

Lemma 7.11.1 *Let a bounded operator A in H have the representation*

$$A = \sum_{k=1}^{n} \phi_k \Delta P_k + V \ (\Delta P_k = P_k - P_{k-1}),$$

where ϕ_k $(k = 1, ..., n)$ are numbers and V is a Hilbert-Schmidt operator satisfying the relations

$$P_{k-1}VP_k = VP_k \quad (k = 1, ..., n).$$

In addition, let f be holomorphic on a neighborhood of the closed convex hull $co(A)$ of $\sigma(A)$. Then

$$\|f(A)\| \leq \sum_{k=0}^{\infty} \sup_{\lambda \in co(A)} |f^{(k)}(\lambda)| \frac{N_2^k(V)}{(k!)^{3/2}}.$$

Proof: Put

$$D = \sum_{k=1}^{n} \phi_k \Delta P_k.$$

Clearly, the spectrum of D consists of numbers ϕ_k $(k = 1, ..., n)$. It is simple to check that $V^n = 0$. That is, V is a nilpotent operator. Due to Lemma 7.5.1, $\sigma(D) = \sigma(A)$. Consequently, ϕ_k $(k = 1, ..., n)$ are eigenvalues of A. Furthermore, let $\{e_m^{(k)}\}_{m=1}^{\infty}$ be an orthogonal normal basis in $\Delta P_k H$. Put

$$Q_l^{(k)} = \sum_{m=1}^{l} (., e_m^{(k)}) e_m^{(k)} \quad (k = 1, ..., n; \ l = 1, 2, ...).$$

Clearly, $Q_l^{(k)}$ strongly converge to ΔP_k as $l \to \infty$. Moreover,

$$Q_l^{(k)} \Delta P_k = \Delta P_k Q_l^{(k)} = Q_l^{(k)}.$$

Then the operators

$$D_l = \sum_{k=1}^{n} \phi_k Q_l^{(k)}$$

strongly tend to D as $l \to \infty$. We can write out,

$$V = \sum_{k=1}^{n} \sum_{i=1}^{k-1} \Delta P_i V \Delta P_k.$$

Introduce the operators

$$W_l = \sum_{k=1}^{n} \sum_{i=1}^{k-1} Q_l^{(i)} V Q_l^{(k)}.$$

Since projectors $Q_l^{(k)}$ strongly converge to ΔP_k as $l \to \infty$, and V is compact, operators W_l converge to V in the operator norm. So the finite dimensional operators

$$T_l := D_l + W_l$$

strongly converge to A and $f(T_l)$ strongly converge to $f(A)$. Thus,

$$\|f(A)\| \leq \sup_l \|f(T_l)\|, \tag{11.1}$$

thanks to the Banach-Steinhaus theorem (see e.g. (Dunford and Schwartz, 1966)). But W_l are nilpotent, and W_l and D_l have the same invariant subspaces. Consequently, due to Lemma 7.5.1,

$$\sigma(D_l) = \sigma(T_l) \subseteq \sigma(A) = \{\phi_k\}.$$

The dimension of T_l is nl. Due to Lemma 2.8.2 and Corollary 6.9.2, we have the inequality

$$\|f(T_l)\| \leq \sum_{k=0}^{ln-1} \sup_{\lambda \in co(A)} |f^{(k)}(\lambda)| \frac{N_2^k(W_l)}{(k!)^{3/2}}.$$

Letting $l \to \infty$ in this inequality, we prove the stated result. $\square$

Lemma 7.11.2 *Let A be a bounded P-triangular operator, whose nilpotent part $V \in C_2$. Let f be a function holomorphic on a neighborhood of $co(A)$. Then*

$$\|f(A)\| \leq \sum_{k=0}^{\infty} \sup_{\lambda \in co(A)} |f^{(k)}(\lambda)| \frac{N_2^k(V)}{(k!)^{3/2}}.$$

Proof: Let D be the diagonal part of A. According to (2.4) and the von Neumann Theorem (Ahiezer and Glazman, 1981, Section 92), there exists a bounded measurable function ϕ, such that

$$D = \int_a^b \phi(t) dP(t).$$

Define the operators

$$V_n = \sum_{k=0}^{n} P(t_{k-1}) V \Delta P_k \text{ and } D_n = \sum_{k=1}^{n} \phi(t_k) \Delta P_k$$

$(t_k = t_k^{(n)}, \ a = t_0 \leq t_1 \leq ... \leq t_n = b; \ \Delta P_k = P(t_k) - P(t_{k-1}), \ k = 1, ..., n)$. Besides, put $B_n = D_n + V_n$. Then the sequence $\{B_n\}$ strongly converges to A due to the triangular representation (2.2). According to (10.1) the sequence $\{f(B_n)\}$ strongly converges to $f(A)$. The inequality

$$\|f(A)\| \leq \sup_n \|f(B_n)\| \tag{11.2}$$

is true thanks to the Banach-Steinhaus theorem. Since the spectral resolution of B_n consists of $n < \infty$ projectors, Lemma 7.11.1 yields the inequality

$$\|f(B_n)\| \leq \sum_{k=0}^{\infty} \sup_{\lambda \in co(B_n)} |f^{(k)}(\lambda)| \frac{N_2^k(V_n)}{(k!)^{3/2}}. \tag{11.3}$$

Thanks to Lemma 7.5.1 we have $\sigma(B_n) = \sigma(D_n)$. Clearly, $\sigma(D_n) \subseteq \sigma(D)$. Hence,

$$\sigma(B_n) \subseteq \sigma(A). \tag{11.4}$$

Due to the well-known Theorem III.7.1 from (Gohberg and Krein, 1970), $\{N_2(V_n)\}$ tends to $N_2(V)$ as n tends to infinity. Thus (11.2), (11.3) and (11.4) imply the required result. $\square$

The assertion of Theorem 7.10.1 immediately follows from Lemmas 7.11.2 and 7.7.2. $\square$

7.12 Regular Functions of Unbounded Operators

Let an operator A be unbounded with a dense domain $Dom\ (A) = Dom\ (A^*)$. In addition, let the conditions (10.2) and

$$\beta(A) := \inf\ Re\ \sigma(A) > -\infty \tag{12.1}$$

hold. Let f be analytic on some neighborhood of $\sigma(A)$ and M an open set containing $\sigma(A)$ whose boundary C consists of a finite number of Jordan arcs such that f is analytic on $M \cup C$. Let C have positive orientation with respect to M. Then we define the function of the operator A by the equality

$$f(A) = f(\infty)I - \frac{1}{2\pi i} \int_C f(\lambda)R_\lambda(A)d\lambda, \tag{12.2}$$

cf. (Dunford and Schwartz, 1966, p. 601).

Without loss of generality we assume that $f(\infty) = 0$. In the other case we can consider the function $f(\lambda) - f(\infty)$.

Theorem 7.12.1 *Under the conditions $D(A) = D(A^*)$, (10.2) and (12.1), let f be regular on a neighborhood of $co(A)$ and $f(\infty) = 0$. Then inequality (10.3) is true.*

Proof: Due to Theorem 7.6.1, under (10.2), A admits the triangular representation. According to (2.4) and the von Neumann Theorem (Ahiezer and Glazman, 1981, Section 92), there exists a P-measurable scalar function ϕ, such that

$$D = \int_{-\infty}^{\infty} \phi(t)dP(t).$$

In addition, as it was shown in the proof of Theorem 7.6.1, the function $Re\ \phi(t)$ nondecreases as t increases. Thus due to Lemma 7.5.1,

$$\inf_t\ Re\ \phi(t) = \beta(D) = \beta(A) > -\infty \text{ and } \sup_t\ |Im\ \phi(t)| < \infty.$$

So the operators $A_n = AP(n)$ are bounded. Moreover, relations $\sigma(A_n) \subseteq \sigma(A)$ and

$$(I\lambda - A)^{-1}P(n) = (I\lambda - A_n)^{-1}P(n) \tag{12.3}$$

hold. Hence, due to (12.2)

$$f(A)P(n) = \frac{1}{2\pi i} \int_C f(\lambda)(I\lambda - A_n)^{-1}d\lambda P(n) = f(A_n)P(n).$$

Due to Theorem 7.10.1,

$$\|f(A_n)\| \le \sum_{k=0}^{\infty} \sup_{\lambda \in co(A_n)} |f^{(k)}(\lambda)| \frac{g_I^k(A_n)}{(k!)^{3/2}}.$$

Letting in this relation $n \to \infty$, we get inequality (10.3). Thus, the theorem is proved. $\square$

Theorem 12.1 is exact. Indeed, under (12.1), let A be normal and (10.5) hold. Then we have equality (10.4), provided $f(\infty) = 0$.

Furthermore, under conditions (12.1) and (10.2), put

$$e^{-At} := \frac{1}{2\pi i} \int_{c_0-i\infty}^{c_0+i\infty} e^{t\lambda}(I\lambda + A)^{-1}d\lambda \quad (c_0 > -\beta(A)). \tag{12.4}$$

Since the non-real spectrum of A is bounded, the integral in (12.4) converges in the sense of the Laplace transformation. Clearly, function e^{zt} is non-analytic at infinity. Let $A_n = AP(n)$. According to (12.4) and (12.3),

$$e^{-At}P(n) = \frac{1}{2\pi i} \int_{c_0-i\infty}^{c_0+i\infty} e^{t\lambda}(I\lambda + A_n)^{-1}P(n)d\lambda = e^{-A_n t}P(n).$$

Due to Theorem 7.10.1

$$\|exp\,[-A_n t]\| \le e^{-t\beta(A_n)} \sum_{k=0}^{\infty} \frac{t^k g_I^k(A_n)}{(k!)^{3/2}},$$

since A_n is bounded. Letting in this relation $n \to \infty$, we get

Theorem 7.12.2 *Let conditions (10.2) and (12.1) hold. Then*

$$\|exp\,[-At]\| \le e^{-t\beta(A)} \sum_{k=0}^{\infty} \frac{t^k g_I^k(A)}{(k!)^{3/2}} \quad (t \ge 0).$$

This theorem is exact. Indeed, let A be normal. Then we have

$$\|exp\,[-At]\| = e^{-t\beta(A)}, \; t \ge 0.$$

For a scalar-valued function h defined on $[0, \infty)$, let the integral

$$\Phi(A) = \int_0^\infty e^{-At} h(t) dt$$

strongly converges. Denote,

$$\theta_k(\Phi, A) := \frac{g_I^k(A)}{(k!)^{3/2}} \int_0^\infty e^{-t\beta(A)} |h(t)| t^k dt \quad (k = 0, 1, ...).$$

Then due to Theorem 7.12.2,

$$\|\Phi(A)\| \le \sum_{k=0}^\infty \theta_k(\Phi, A),$$

provided the integrals and series converge. In particular, let

$$\beta(A) > 0. \tag{12.5}$$

Then

$$A^{-m} = \frac{1}{(m-1)!} \int_0^\infty e^{-At} t^{m-1} dt \quad (m = 1, 2, ...).$$

In this case

$$\theta_k(\Phi, A) = \frac{g_I^k(A)}{(m-1)!(k!)^{3/2}} \int_0^\infty e^{-t\beta(A)} t^{m+k-1} dt.$$

Therefore

$$\|A^{-m}\| \le \sum_{k=0}^\infty \frac{(k+m-1)! g_I^k(A)}{(m-1)!(k!)^{3/2} \beta(A)^{m+k}}. \tag{12.6}$$

In addition, under (12.5)

$$A^{-\nu} = \frac{1}{\Gamma(\nu)} \int_0^\infty e^{-At} t^{\nu-1} dt \quad (0 < \nu < 1),$$

where $\Gamma(.)$ is the Euler Gamma-function. Hence,

$$\|A^{-\nu}\| \le \sum_{k=0}^\infty \frac{\Gamma(\nu+k) g_I^k(A)}{\Gamma(\nu)(k!)^{3/2} \beta(A)^{k+\nu}}. \tag{12.7}$$

7.13 Triangular Representations of Regular Functions

Lemma 7.13.1 *Let a bounded operator A admit a triangular representation with some maximal resolution of the identity. In addition, let f be a function analytic on a neighborhood of $\sigma(A)$. Then the operator $f(A)$ admits the triangular representation with the same maximal resolution of the identity. Moreover, the diagonal part D_f of $f(A)$ is defined by $D_f = f(D)$, where D is the diagonal part of A.*

Proof: Due to representations (10.1) and (2.2),

$$f(A) = -\frac{1}{2\pi i} \int_C f(\lambda) R_\lambda(A) d\lambda = -\frac{1}{2\pi i} \int_C f(\lambda) R_\lambda(D)(I + V R_\lambda(D))^{-1} d\lambda.$$

Consequently,

$$f(A) = -\frac{1}{2\pi i} \int_C f(\lambda) R_\lambda(D) d\lambda + W = f(D) + W, \qquad (13.1)$$

where

$$W = -\frac{1}{2\pi i} \int_C f(\lambda) R_\lambda(D)[(I + V R_\lambda(D))^{-1} - I] d\lambda.$$

But

$$(I + V R_\lambda(D))^{-1} - I = -V R_\lambda(D)(I + V R_\lambda(D))^{-1} \quad (\lambda \in C).$$

We thus get

$$W = \frac{1}{2\pi i} \int_C f(\lambda)\psi(\lambda) d\lambda,$$

where

$$\psi(\lambda) = R_\lambda(D) V R_\lambda(D)(I + V R_\lambda(D))^{-1}.$$

Let $P(.)$ be a maximal resolution of the identity of A. Due to Lemma 7.3.3, for each $\lambda \in C$, $\psi(\lambda)$ is a Volterra operator with the same m.r.i. Since $P(t)$ is a bounded operator, we have by Lemma 7.3.2

$$(P(t_0 + 0) - P(t_0))W(P(t_0 + 0) - P(t_0)) =$$

$$\frac{1}{2\pi i} \int_C f(\lambda)(P(t_0 + 0) - P(t_0))\psi(\lambda)(P(t_0 + 0) - P(t_0)) d\lambda = 0$$

for every gap $P(t_0 + 0) - P(t_0)$ of $P(t)$. Thus, W is a Volterra operator thanks to Lemma 7.3.1. This and (13.1) prove the lemma. □

7.14   Triangular Representations of Quasiunitary Operators

A bounded linear operator A is called *a quasiunitary operator*, if $A^*A - I$ is a completely continuous operator.

Lemma 7.14.1 *Let A be a linear operator in H satisfying the condition*

$$A^*A - I \in C_p \quad (1 \le p < \infty), \qquad (14.1)$$

and let the operator $I - A$ be invertible. Then the operator

$$B = i(I - A)^{-1}(I + A) \qquad (14.2)$$

(Cayley's transformation of A) is bounded and satisfies the condition $B - B^* \in C_p$.

Proof: According to (14.1) we can write down

$$A = U(I + K_0) = U + K,$$

where U is a unitary operator and both operators K and K_0 belong to C_p and $K_0 = K_0^*$. Consequently,

$$B^* = -i(I + U^* + K^*)(I - U^* - K^*)^{-1} =$$

$$-i(I + U^{-1} + K^*)(I - U^{-1} - K^*)^{-1}.$$

That is, $B^* = -i(I + U + K_0)(U - I - K_0)^{-1}$, since $K_0 = K^*U$. But (14.2) clearly forces

$$B = i(I + U + K)(I - U - K)^{-1}.$$

Thus, $2B_I = (B - B^*)/i = T_1 + T_2$, where

$$T_1 = (I + U)[(I - U - K)^{-1} + (U - I - K_0)^{-1}]$$

and

$$T_2 = K(I - U - K)^{-1} + K_0(U - I - K_0)^{-1}.$$

Since $K, K_0 \in C_p$, we conclude that $T_2 \in C_p$. It remains to prove that $T_1 \in C_p$. Let us apply the identity

$$(I - U - K)^{-1} + (U - I - K_0)^{-1} = -(I - U - K)^{-1}(K_0 + K)(U - I - K_0)^{-1}.$$

Hence, $T_1 \in C_p$. This completes the proof. $\square$

Lemma 7.14.2 *Under condition (14.1), let A have a regular point on the unit circle. Then A admits the triangular representation (2.2)-(2.4).*

Proof: Without any loss of generality we assume that A has on the unit circle a regular point $\lambda_0 = 1$. In the other case we can consider instead of A the operator $A\lambda_0^{-1}$.

Let us consider in H the operator B defined by (14.2). By the previous lemma it satisfies (14.2) and therefore due to Theorem 7.6.1 has a triangular representation. The transformation inverse to (14.2) must be defined by the formula

$$A = (B - iI)(B + iI)^{-1}. \tag{14.3}$$

Now Lemma 7.13.1 ensures the result. $\square$

7.15 Resolvents and Analytic Functions of Quasiunitary Operators

Assume that A has a regular point on the unit circle and

$$AA^* - I \text{ is a nuclear operator.} \tag{15.1}$$

Due to Lemma 7.14.1, from (1.1), (14.3) and (15.1) it follows that

$$\sum_{k=1}^{\infty}(|\lambda_k(A)|^2 - 1) < \infty,$$

where $\lambda_k(A), k = 1, 2, ...$ are the nonunitary eigenvalues with their multiplicities, that is, the eigenvalues with the property $|\lambda_k(A)| \neq 1$. Under (15.1) put

$$\vartheta(A) = [Tr \ (A^*A - I) - \sum_{k=1}^{\infty}(|\lambda_k(A)|^2 - 1)]^{1/2}.$$

If A is a normal operator, then $\vartheta(A) = 0$. Let A have the unitary spectrum, only. That is, $\sigma(A)$ lies on the unit circle. Then

$$\vartheta(A) = [Tr \ (A^*A - I)]^{1/2}.$$

Moreover, if the condition

$$\sum_{k=1}^{\infty}(|\lambda_k(A)|^2 - 1) \geq 0 \qquad (15.2)$$

holds, then

$$Tr \ (A^*A - I) = \sum_{k=1}^{\infty}(s_k^2(A) - 1) \geq \sum_{k=1}^{\infty}(|\lambda_k(A)|^2 - 1) = Tr \ (D^*D - I) \geq 0$$

and therefore, under (15.2),

$$\vartheta(A) \leq [Tr \ (A^*A - I)]^{1/2}. \qquad (15.3)$$

Theorem 7.15.1 *Under condition (15.1), let an operator A have a regular point on the unit circle. Then*

$$\|R_\lambda(A)\| \leq \sum_{k=0}^{\infty} \frac{\vartheta^k(A)}{\sqrt{k!}\rho^{k+1}(A, \lambda)} \quad (\lambda \notin \sigma(A)). \qquad (15.4)$$

Moreover, there are constants $a_0, b_0 > 0$, independent of λ, such that

$$\|R_\lambda(A)\| \leq \frac{a_0}{\rho(A, \lambda)} exp \ [\frac{b_0 \ \vartheta^2(A)}{\rho^2(A, \lambda)}] \quad (\lambda \notin \sigma(A)). \qquad (15.5)$$

These constants can be taken from (5.10) or from (5.11).

To prove this theorem we need the following

Lemma 7.15.2 *Under condition (15.1), let an operator A have a regular point on the unit circle. Then A admits the triangular representation (2.2) (due to Lemma 7.14.2). Moreover, $N_2(V) = \vartheta(A)$, where V is the nilpotent part of A.*

Proof: By Lemma 7.3.3 and Theorem 6.2.2 $Tr\ (D^*V) = Tr\ (V^*D) = 0$. Employing the triangular representation (2.2) we obtain

$$Tr\ (A^*A - I) = Tr\ [(D+V)^*(D+V) - I)] =$$

$$Tr\ (D^*D - I) + Tr\ (V^*V).$$

Since D is a normal operator and the spectra of D and A coincide due to Lemma 7.5.1, then we can write

$$Tr\ (D^*D - I) = \sum_{k=1}^{\infty} (|\lambda_k(A)|^2 - 1).$$

This equality implies the required result. $\square$

Proof of Theorem 7.15.1:   Inequality (15.4) follows from relation (5.7) and Lemma 7.15.2. Inequality (15.5) follows from Theorem 7.5.5 and Lemma 7.15.2. $\square$

Let us extend Theorem 7.15.1 to the case

$$A^p(A^*)^p - I \in C_1 \tag{15.6}$$

for some integer number $p > 1$.

Corollary 7.15.3 *Under condition (15.6), let an operator A have a regular point on the unit circle. Then*

$$\|R_\lambda(A)\| \leq \|T_{\lambda,p}\| \sum_{k=0}^{\infty} \frac{\vartheta^k(A^p)}{\rho^{k+1}(A^p, \lambda^p)\sqrt{k!}} \quad (\lambda^p \notin \sigma(A^p)),$$

where

$$T_{\lambda,p} = \sum_{k=0}^{p-1} A^k \lambda^{p-k-1} \ \text{and} \ \rho(A^p, \lambda^p) = \inf_{t\in\sigma(A)} |t^p - \lambda^p|.$$

Moreover, there are constants $a_0, b_0 > 0$, independent of λ, such that

$$\|R_\lambda(A)\| \leq \frac{a_0\|T_{\lambda,p}\|}{\rho(A^p, \lambda^p)} \ exp\ [\frac{b_0\ \vartheta^2(A^p)}{\rho^2(A^p, \lambda^p)}] \quad (\lambda^p \notin \sigma(A^p)).$$

These constants can be taken from (5.10) or from (5.11).

This result is due Theorem 7.15.1 and identity (8.3).

Theorem 7.15.4 *Let a linear operator A satisfy condition (15.1) and have a regular point on the unit circle. If, in addition, f is a holomorphic function on a neighborhood of closed convex hull $co(A)$ of $\sigma(A)$, then*

$$\|f(A)\| \leq \sum_{k=0}^{\infty} \sup_{\lambda\in co(A)} |f^{(k)}(\lambda)| \frac{\vartheta^k(A)}{(k!)^{3/2}}. \tag{15.7}$$

Proof: The result immediately follows from Lemmas 7.15.2 and 7.11.2. $\square$

Theorem 7.15.4 is precise: inequality (15.7) becomes equality (10.4) if A is a unitary operator and (10.5) holds, because $\vartheta(A) = 0$ in this case.

Example 7.15.5 *Let an operator A satisfy the condition (15.1) and have a regular point on the unit circle. Then the inequality*

$$\|A^m\| \le \sum_{k=0}^{m} \frac{m! r_s^{m-k}(A)\vartheta^k(A)}{(m-k)!(k!)^{3/2}}$$

holds for any integer $m \ge 1$. Recall that $r_s(A)$ is the spectral radius of A.

7.16   Notes

Notions similar to Definitions 7.1.1, 7.2.1 and 7.2.2 can be found in the books (Gohberg and Krein, 1970) and (Brodskii, 1971), as well as in the papers (Branges, 1963, 1965a and 1965b) and (Brodskii, Gohberg, and Krein, 1969).

Theorem 7.5.3 is probably new.

The results presented in Sections 7.6-7.8 are based on Chapter 3 of the book (Gil', 1995), but the proofs are considerably improved.

Theorem 7.9.1 was established in the paper (Gil', 1993). Theorems 7.10.1 and 7.12.1 were derived in the paper (Gil', 1992).

Triangular representations of quasi-Hermitian and quasiunitary operators can be found, in particular, in the paper by V. Brodskii, I. Gohberg and M. Krein (1969), and references given therein.

References

[1] Ahiezer, N. I. and Glazman, I. M. (1981). *Theory of Linear Operators in a Hilbert Space.* Pitman Advanced Publishing Program, Boston, London, Melburn.

[2] Branges, L. de. (1963). Some Hilbert spaces of analytic functions I, *Proc. Amer. Math. Soc.* **106**, 445-467.

[3] Branges, L. de. (1965a). Some Hilbert spaces of analytic functions II, *J. Math. Analysis and Appl.*, **11**, 44-72.

[4] Branges, L. de. (1965b). Some Hilbert spaces of analytic functions III, *J. Math. Analysis and Appl.*, **12**, 149-186.

[5] Brodskii, M. S. (1971). *Triangular and Jordan Representations of Linear Operators*, Transl. Math. Mongr., v. 32, Amer. Math. Soc., Providence, R.I.

[6] Brodskii, V.M., Gohberg, I.C. and Krein M.G. (1969). General theorems on triangular representations of linear operators and multiplicative representations of their characteristic functions, *Funk. Anal. i Pril.*, **3**,1-27 (in Russian); English Transl., *Func. Anal. Appl.* **3**, 255-276.

[7] Dunford, N and Schwartz, J. T. (1966). *Linear Operators, part I. General Theory*. Interscience publishers, New York, London.

[8] Dunford, N and Schwartz, J. T. (1963). *Linear Operators, part II. Spectral Theory*. Interscience publishers, New York, London.

[9] Gil , M. I. (1992). One estimate for the norm of a function of a quasihermitian operator, *Studia Mathematica*, **103(1)**, 17-24.

[10] Gil', M. I. (1993). Estimates for Norm of matrix-valued and operator-value functions, *Acta Applicandae Mathematicae* **32**, 59-87.

[11] Gil', M. I. (1995). *Norm Estimations for Operator-Valued Functions and Applications*. Marcel Dekker, Inc, New York.

[12] Gohberg, I. C. and Krein, M. G. (1969). *Introduction to the Theory of Linear Nonselfadjoint Operators*, Trans. Mathem. Monographs, v. 18, Amer. Math. Soc., R.I.

[13] Gohberg, I. C. and Krein, M. G. (1970). *Theory and Applications of Volterra Operators in Hilbert Space*, Trans. Mathem. Monographs, v. 24, Amer. Math. Soc., Providence, R. I.

8. Bounded Perturbations of Nonselfadjoint Operators

In the present chapter we consider the operators of the kind $A + B$, where A is a P-triangular operator and B is a bounded operator. We investigate the invertibility conditions and bounds for the spectra of such operators. In particular, we consider perturbations of the von Neumann - Schatten operators and operators with von Neumann - Schatten Hermitian components.

8.1 Invertibility of Boundedly Perturbed P-Triangular Operators

Throughout the present chapter, A is a P-triangular operator in a separable Hilbert space H with the nilpotent part V and diagonal part D. According to Definition 7.2.2 and Lemma 7.5.1 this means that

$$A = D + V \text{ and } \sigma(D) = \sigma(A). \tag{1.1}$$

Besides, D is a normal operator and V is a Volterra one. Moreover, D and V have the same maximal resolution of the identity. In addition, assume that

$$B \text{ is a bounded linear operator in } H. \tag{1.2}$$

In this section we suppose that D is boundedly invertible:

$$r_l(D) = \inf |\sigma(D)| > 0. \tag{1.3}$$

Then due to Lemma 7.3.3, the operator

$$W := D^{-1}V$$

is a Volterra one. Let V belong to a norm ideal Y with a norm $|\cdot|_Y$, introduced in Section 7.4. Namely, there are positive numbers θ_k $(k \in \mathbf{N})$, with

$$\theta_k^{1/k} \to 0 \text{ as } k \to \infty,$$

such that

$$\|V^k\| \leq \theta_k |V|_Y^k. \tag{1.4}$$

Put

$$J_Y(W) = \sum_{k=0}^{ni(W)-1} \theta_k |W|_Y^k.$$

Recall that $ni(W)$ is the nilpotency index of W. Clearly,

$$J_Y(W) \leq \sum_{k=0}^{\infty} \frac{\theta_k |V|_Y^k}{r_l^{k+1}(D)}.$$

Theorem 8.1.1 *Under conditions (1.1)-(1.4), let*

$$r_l^{-1}(D)\|B\|J_Y(W) < 1.$$

Then the operator $A + B$ is invertible. Moreover,

$$\|(A+B)^{-1}\| \leq \frac{J_Y(W)}{r_l(D) - \|B\|J_Y(W)}.$$

To prove this theorem we need the following simple result

Lemma 8.1.2 *Let A_1, A_2 be linear operators in H. In addition, let A_1 be invertible and*

$$\|B_1 A_1^{-1}\| < 1,$$

where $B_1 = A_2 - A_1$. Then A_2 is also invertible, with

$$\|A_2^{-1}\| \leq \frac{\|A_1^{-1}\|}{1 - \|B_1 A_1^{-1}\|}.$$

Proof: Clearly $A_2 = A_1 + B_1 = (I + B_1 A_1^{-1})A_1$. Hence

$$A_2^{-1} = A_1^{-1} \sum_{k=0}^{\infty} (-A_1^{-1}B_1)^k.$$

This proves the result. $\square$

Proof of Theorem 8.1.1: From (1.1) and (1.4) it follows

$$A^{-1} = (I + D^{-1}V)^{-1}D^{-1} = \sum_{k=0}^{ni(W)-1} W^k D^{-1}$$

So $\|A^{-1}\| \leq r_l^{-1}(D)J_Y(W)$. Now the required result is due to the previous lemma. $\square$

Assume now that the nilpotent part of A belongs to a Neumann-Schatten ideal:

$$N_{2p}(V) := [Trace \ (V^*V)^p]^{1/2p} < \infty \qquad (1.5)$$

for some integer $p \geq 1$. Put

$$J_p(W) = \sum_{k=0}^{ni(W)-1} \theta_k^{(p)} |W|_Y^k,$$

where $\theta_k^{(p)}$ are defined in Section 6.7. Lemma 8.1.2 and Corollary 7.5.4 imply

Corollary 8.1.3 *Under conditions (1.1)-(1.3) and (1.5), let*

$$\|B\|J_p(W) < r_l(D).$$

Then operator $A + B$ is invertible. Moreover,

$$\|(A + B)^{-1}\| \leq \frac{J_p(W)}{r_l(D) - \|B\|J_p(W)}.$$

Note that Theorem 7.5.5 implies under (1.5) the inequality

$$\|A^{-1}\| \leq \psi_p(V, r_l(D)), \qquad (1.6)$$

where

$$\psi_p(V, r_l(D)) = a_0 \sum_{j=0}^{p-1} \frac{N_{2p}^j(V)}{r_l^{j+1}(D)} \ exp \ [\frac{b_0 \ N_{2p}^{2p}(V)}{r_l^{2p}(D)}]. \qquad (1.7)$$

Besides, the constants a_0, b_0 can be taken as in the relations

$$a_0 = \sqrt{\frac{c}{c-1}} \ \text{and} \ b_0 = \frac{c}{2} \ \text{for any} \ c > 1, \ \text{in particular,} \ a_0 = \sqrt{2} \ \text{and} \ b_0 = 1,$$

$$(1.8)$$

or as in the relations

$$a_0 = e^{1/2} \ \text{and} \ b_0 = 1/2. \qquad (1.9)$$

Lemma 8.1.2 and (1.6) yield

Corollary 8.1.4 *Under the conditions (1.1)-(1.3) and (1.5), let*

$$\|B\|\psi_p(V, r_l(D)) < 1.$$

Then operator $A + B$ is invertible. Moreover,

$$\|(A + B)^{-1}\| \le \frac{\psi_p(V, r_l(D))}{1 - \|B\|\psi_p(V, r_l(D))}.$$

8.2   Resolvents of Boundedly Perturbed P-Triangular Operators

We need the following result, which immediately follows from Lemma 8.1.2:

Lemma 8.2.1 *Let A_1, A_2 be linear operators in H. In addition, let λ be a regular point of A_1 and*

$$\|B_1(A_1 - \lambda I)^{-1}\| < 1 \quad (B_1 = A_1 - A_2).$$

Then λ is regular also for A_2, and

$$\|(A_2 - \lambda I)^{-1}\| \le \frac{\|(A_1 - \lambda I)^{-1}\|}{1 - \|B_1(A_1 - \lambda I)^{-1}\|}.$$

Again put $\nu(\lambda) = ni \; (VR_\lambda(D))$. Recall that *one can replace $\nu(\lambda)$ by ∞*. Under (1.4), let

$$J_Y(V, m, z) = \sum_{k=0}^{m-1} \frac{\theta_k |V|_Y^k}{z^{k+1}} \quad (z > 0).$$

Then the previous lemma and Theorem 7.5.3 imply

Theorem 8.2.2 *Under conditions (1.1), (1.2), (1.4), let*

$$\|B\|J_Y(V, \nu(\lambda), \rho(D, \lambda)) < 1.$$

Then λ is a regular point of $A + B$. Moreover,

$$\|(A + B - \lambda I)^{-1}\| \le \frac{J_Y(V, \nu(\lambda), \rho(D, \lambda))}{1 - \|B\|J_Y(V, \nu(\lambda), \rho(D, \lambda))}.$$

Under (1.5) denote

$$\tilde{J}_p(V, m, z) = \sum_{k=0}^{m-1} \frac{N_{2p}^k(V)}{z^{k+1}} \quad (z > 0).$$

Then Lemma 8.2.1 and Corollary 7.5.4 yield

Corollary 8.2.3 *Under conditions (1.1), (1.2) and (1.5), let*

$$\|B\|\tilde{J}_p(V, \nu(\lambda), \rho(D, \lambda)) < 1.$$

Then λ is a regular point of $A + B$. Moreover,

$$\|(A + B - \lambda I)^{-1}\| \leq \frac{\tilde{J}_p(V, \nu(\lambda), \rho(D, \lambda))}{1 - \|B\|\tilde{J}_p(V, \nu(\lambda), \rho(D, \lambda))}.$$

Furthermore, put

$$\psi_p(V, z) = a_0 \sum_{k=0}^{p-1} \frac{N_{2p}^k(V)}{z^{k+1}} \exp\left[\frac{b_0 N_{2p}^{2p}(V)}{z^{2p}}\right] \quad (z > 0, V \in C_{2p}),$$

where a_0, b_0 do not depend on z and can be taken as in (1.8) or as in (1.9). Now Theorem 7.5.5 and Lemma 8.2.1 yield

Corollary 8.2.4 *Under conditions (1.1), (1.2) and (1.5), let*

$$\|B\|\psi_p(V, \rho(D, \lambda)) < 1.$$

Then λ is a regular point of $A + B$. Moreover,

$$\|(A + B - \lambda I)^{-1}\| \leq \frac{\psi_p(V, \rho(D, \lambda))}{1 - \|B\|\psi_p(V, \rho(D, \lambda))}.$$

8.3   Roots of Scalar Equations

Consider the scalar equation

$$\sum_{k=1}^{\infty} a_k z^k = 1 \tag{3.1}$$

where the coefficients $a_k, k = 1, 2, \ldots$ have the property

$$\gamma_0 \equiv 2 \max_k \sqrt[k]{|a_k|} < \infty.$$

We will need the following

Lemma 8.3.1 *Any root z_0 of equation (3.1) satisfies the estimate $|z_0| \geq 1/\gamma_0$.*

Proof: Set in (3.1) $z_0 = x\gamma_0^{-1}$. We have

$$1 = \sum_{k=1}^{\infty} a_k \gamma_0^{-k} x^k \leq \sum_{k=1}^{\infty} |a_k| \gamma_0^{-k} |x|^k. \tag{3.2}$$

But

$$\sum_{k=1}^{\infty} |a_k| \gamma_0^{-k} \le \sum_{k=1}^{\infty} 2^{-k} = 1$$

and therefore, $|x| \ge 1$. Hence, $|z_0| = \gamma_0^{-1}|x| \ge \gamma_0^{-1}$. As claimed. $\square$

Note that the latter lemma generalizes the well-known result for algebraic equations, cf. the book (Ostrowski, 1973, p. 277).

Lemma 8.3.2 *The extreme right (unique positive) root z_a of the equation*

$$\sum_{j=0}^{p-1} \frac{1}{y^{j+1}} \ exp \ [\frac{1}{2}(1 + \frac{1}{y^{2p}})] = a \ \ (a \equiv const > 0) \tag{3.3}$$

satisfies the inequality $z_a \le \delta_p(a)$, where

$$\delta_p(a) := \begin{cases} pe/a & \text{if } a \le pe, \\ [ln \ (a/p)]^{-1/2p} & \text{if } a > pe \end{cases} \tag{3.4}$$

Proof: Assume that

$$pe \ge a. \tag{3.5}$$

Since the function

$$f(y) \equiv \sum_{j=0}^{p-1} \frac{1}{y^{j+1}} \ exp \ [\frac{1}{2}(1 + \frac{1}{y^{2p}})]$$

is nonincreasing and $f(1) = pe$, we have $z_a \ge 1$. But due to (3.3),

$$z_a = 1/a \sum_{j=0}^{p-1} z_a^{-j} exp \ [(1 + z_a^{-2p})/2] \le pe/a.$$

So in the case (3.5), the lemma is proved. Let now

$$pe < a. \tag{3.6}$$

Then $z_a \le 1$. But

$$\sum_{j=0}^{p-1} x^{j+1} \le px^p \le p \ exp \ [x^p - 1] \le p \ exp \ [(x^{2p} + 1)/2 - 1]$$

$$= p \ exp \ [x^{2p}/2 - 1/2] \ \ (x \ge 1).$$

So

$$f(y) = \sum_{j=0}^{p-1} \frac{1}{y^{j+1}} \ exp \ [\frac{1}{2}(1 + \frac{1}{y^{2p}})] \le p \ exp \ [\frac{1}{y^{2p}}] \ \ (y \le 1).$$

But $z_a \leq 1$ under (3.6). We thus have

$$a = f(z_a) \leq p \; exp \; [\frac{1}{z_a^{2p}}].$$

Or

$$z_a^{2p} \leq 1/ln \; (a/p).$$

This finishes the proof. $\square$

8.4   Spectral Variations

Definition 8.4.1 *Let A and B be linear operators in H. Then the quantity*

$$sv_A(B) := \sup_{\mu \in \sigma(B)} \; \inf_{\lambda \in \sigma(A)} |\mu - \lambda|$$

is called the spectral variation of a B with respect to A. In addition,

$$hd(A, B) := \max\{sv_A(B), sv_B(A)\}$$

is the Hausdorff distance between the spectra of A and B.

First, we will prove the following technical lemma

Lemma 8.4.2 *Let A_1 and A_2 be linear operators in H with the same domain and*

$$q \equiv \|A_1 - A_2\| < \infty.$$

In addition, let

$$\|R_\lambda(A_1)\| \leq F(\rho^{-1}(A_1, \lambda)) \; (\lambda \notin \sigma(A_1)) \tag{4.1}$$

where $F(x)$ is a monotonically increasing non-negative function of a non-negative variable x, such that $F(0) = 0$ and $F(\infty) = \infty$. Then

$$sv_{A_1}(A_2) \leq z(F, q),$$

where $z(F, q)$ is the extreme right-hand (positive) root of the equation

$$1 = qF(1/z). \tag{4.2}$$

Proof: Due to (4.1) and Lemma 8.2.1,

$$1 \leq qF(\rho^{-1}(A, \mu)) \text{ for all } \mu \in \sigma(B). \tag{4.3}$$

Compare this inequality with (4.2). Since $F(x)$ monotonically increases, $z(F, q)$ is a unique positive root of (4.2), and $\rho(A, \mu) \leq z(F, q)$. This proves the required result. $\square$

The previous lemma and Theorem 7.5.3 imply

Theorem 8.4.3 *Let conditions (1.1), (1.2) and (1.4) hold. Then*

$$sv_D(A+B) \le z_Y(V,B)$$

where $z_Y(V,B)$ is the extreme right-hand (positive) root of the equation

$$1 = \|B\| J_Y(\nu_0, V, z) \equiv \|B\| \sum_{k=0}^{\nu_0-1} \frac{\theta_k |V|_Y^k}{z^{k+1}}$$

and

$$\nu_0 = \sup_{\lambda \notin \sigma(D)} ni\,(VR_\lambda(D)).$$

Note that to estimate the root $z_Y(V,B)$, one can use Lemma 8.3.1.

Moreover, Lemma 8.4.2 and Theorem 7.5.5 imply

Theorem 8.4.4 *Let conditions (1.1), (1.2) and (1.5) hold. Then $sv_D(A+B)$ $\le z_p(V,B)$, where $z_p(V,B)$ is the extreme right-hand (positive) root of the equation*

$$1 = \|B\| \sum_{j=0}^{p-1} \frac{N_{2p}^j(V)}{z^{j+1}} \, exp\,[\frac{1}{2} + \frac{N_{2p}^{2p}(V)}{2z^{2p}}]. \tag{4.4}$$

Substitute in (4.4) the equality $z = xN_{2p}(V)$. Then we have equation (3.3) with

$$a = \frac{N_{2p}(V)}{\|B\|}.$$

Thanks to Lemma 8.3.2, we get $y_p(V,B) \le \delta_p(V,B)$, where

$$\delta_p(V,B) := N_{2p}(V)\delta_p(\frac{N_{2p}(V)}{\|B\|})$$

and $\delta_p(.)$ is defined by (3.4). We thus have derived

Corollary 8.4.5 *Under the hypothesis of Theorem 8.4.4, $sv_D(A+B) \le \delta_p(V,B)$.*

8.5   Perturbations of Compact Operators

First assume that

$$A \in C_2. \tag{5.1}$$

By virtue of Lemma 8.4.2 and Theorem 6.4.2 we arrive at the following result.

Theorem 8.5.1 *Let condition (5.1) hold and B be a bounded operator in H. Then*

$$sv_A(A + B) \leq \tilde{z}_1(A, B),$$

where $\tilde{z}_1(A, B)$ is the extreme right-hand (positive) root of the equation

$$1 = \frac{\|B\|}{z} \, exp \, [\frac{1}{2} + \frac{g^2(A)}{2z^2}]. \qquad (5.2)$$

Furthermore, substitute in (5.2) the equality $z = xg(A)$. Then we arrive at the equation

$$\frac{1}{z} \, exp \, [\frac{1}{2} + \frac{1}{2z^2}] = \frac{g(A)}{\|B\|}.$$

Applying Lemma 8.3.2 to this equation, we get $\tilde{z}_1(A, B) \leq \tilde{\Delta}_1(A, B)$, where

$$\tilde{\Delta}_1(A, B) := g(A)\delta_1(g(A)/\|B\|)$$

where δ_1 is defined by (3.4) with $p = 1$. So

$$\tilde{\Delta}_1(A, B) := \begin{cases} e\|B\| & \text{if } g(A) \leq e\|B\|, \\ g(A) \, [ln \, (g(A)/\|B\|)]^{-1/2} & \text{if } g(A) > e\|B\| \end{cases} . \qquad (5.3)$$

Now Theorem 8.5.1 yields

Corollary 8.5.2 *Under the hypothesis of Theorem 8.5.1, for any $\mu \in \sigma(A + B)$, there is a $\mu_0 \in \sigma(A)$, such that $|\mu - \mu_0| \leq \tilde{\delta}_1(A, B)$. In particular,*

$$r_s(A + B) \leq r_s(A) + \tilde{\Delta}_1(A, B), \qquad (5.4)$$

$$r_l(A + B) \geq \max\{r_l(A) - \tilde{\Delta}_1(A, B), 0\} \text{ and } \alpha(A + B) \leq \alpha(A) + \tilde{\Delta}_1(A, B). \qquad (5.5)$$

Remark 8.5.3 *According to Lemma 6.5.2, in Theorem 8.5.1 and its corollary, one can replace $g(A)$ by $\sqrt{2}N_2(A_I)$.*

Now let

$$A \in C_{2p} \quad (p = 2, 3, ...). \qquad (5.6)$$

Then by virtue of Lemma 8.4.2 and Theorem 6.7.3 we arrive at the following result.

Theorem 8.5.4 *Let condition (5.6) hold and B be a bounded operator in H. Then $sv_A(A + B) \leq \tilde{y}_p(A, B)$, where $\tilde{y}_p(A, B)$ is the extreme right-hand (positive) root of the equation*

$$1 = \|B\| \sum_{m=0}^{p-1} \frac{(2N_{2p}(A))^m}{z^{m+1}} \, exp \, [\frac{1}{2} + \frac{(2N_{2p}(A))^{2p}}{2z^{2p}}]. \qquad (5.7)$$

Furthermore, substitute in (5.7) the equality $z = x2N_{2p}(A)$ and apply Lemma 8.3.2. Then we get $\tilde{y}_p(A, B) \leq \Delta_p(A, B)$, where

$$\Delta_p(A, B) := 2N_{2p}(A)\delta_p(2N_{2p}(A)/\|B\|).$$

Recall that $\delta_p(.)$ is defined by (3.4). So

$$\Delta_p(A, B) := \begin{cases} pe\|B\| & \text{if } 2N_{2p}(A) \leq ep\|B\|, \\ 2N_{2p}(A) \left[ln \ (2N_{2p}(A)/p\|B\|)\right]^{-1/2p} & \text{if } 2N_{2p}(A) > ep\|B\| \end{cases}.$$
(5.8)

Now Theorem 8.5.4 yields

Corollary 8.5.5 *Under the hypothesis of Theorem 8.5.4, for any $\mu \in \sigma(A + B)$, there is a $\mu_0 \in \sigma(A)$, such that $|\mu - \mu_0| \leq \Delta_p(A, B)$. In particular, relations (5.4) and (5.5) hold with $\Delta_p(A, B)$ instead of $\tilde{\Delta}_1(A, B)$.*

Remark 8.5.6 *According to Theorem 7.9.1, in Theorem 8.5.4 and its corollary one can replace $2N_{2p}(A)$ by $\tilde{\beta}_p N_{2p}(A_I)$.*

8.6   Perturbations of Operators with Compact Hermitian Components

First, let $Dom \ (A) = Dom \ (A^*)$ and

$$A - A^* \in C_2.$$
(6.1)

Then by virtue of Lemma 8.4.2 and Theorem 7.7.1 we arrive at the following result.

Theorem 8.6.1 *Let condition (6.1) hold and B be a bounded operator in H. Then $sv_A(A + B) \leq x_1(A, B)$, where $x_1(A, B)$ is the extreme right-hand (positive) root of the equation*

$$1 = \|B\| \sum_{k=0}^{\infty} \frac{g_I^k(A)}{\sqrt{k!}z^{k+1}}.$$
(6.2)

Recall that $g_I(A) \leq \sqrt{2}N_2(A_I)$. According to Theorem 7.7.1, one can replace (6.2) by the equation

$$1 = \frac{\|B\|}{z} \ exp \ [\frac{1}{2} + \frac{g_I^2(A)}{2z^2}].$$
(6.3)

Furthermore, substitute in (6.3) the equality $z = xg_I(A)$ and apply Lemma 8.3.2. Then we can assert that extreme right-hand root of equation (6.3) is less than $\tau_1(A, B)$, where

$$\tau_1(A, B) = g_I(A)\delta_1(g_I(A)/\|B\|).$$

So, according to (3.4),

$$\tau_1(A,B) = \begin{cases} e\|B\| & \text{if } g_I(A) \le e\|B\|, \\ g_I(A)\,[ln\,(g_I(A)/\|B\|)]^{-1/2} & \text{if } g_I(A) > e\|B\| \end{cases}.$$

Hence, Theorem 8.6.1 yields

Corollary 8.6.2 *Let condition (6.1) hold and B be a bounded operator in H. Then for any $\mu \in \sigma(A+B)$, there is a $\mu_0 \in \sigma(A)$, such that $|\mu - \mu_0| \le \tau_1(A,B)$. In particular,*

$$r_s(A+B) \le r_s(A) + \tau_1(A,B),$$

$$r_l(A+B) \ge \max\{r_l(A) - \tau_1(A,B), 0\} \text{ and } \alpha(A+B) \le \alpha(A) + \tau_1(A,B).$$

Now let

$$A - A^* \in C_{2p} \quad (p = 2,3,...). \tag{6.4}$$

Then by virtue of Lemma 8.4.2 and Theorem 7.9.1 we arrive at the following result.

Theorem 8.6.3 *Let condition (6.4) hold and B be a bounded operator in H. Then $sv_D(A+B) \le \tilde{x}_p(A,B)$, where $\tilde{x}_p(A,B)$ is the extreme right-hand (positive) root of the equation*

$$1 = \|B\| \sum_{m=0}^{p-1} \sum_{k=0}^{\infty} \frac{w_p^{pk+m}(A)}{\sqrt{k!}\,z^{pk+m+1}}. \tag{6.5}$$

where

$$w_p(A) = \tilde{\beta}_p N_{2p}(A_I).$$

According to Theorem 7.9.1, one can replace (6.5) by the equation

$$1 = \|B\| \sum_{m=0}^{p-1} \frac{(w_p(A))^m}{z^{m+1}}\,exp\,[\frac{1}{2} + \frac{(w_p(A))^{2p}}{2z^{2p}}]. \tag{6.6}$$

Furthermore, substitute in (6.6) the equality $z = xw_p(A)$ and apply Lemma 8.3.2. Then we can assert that the extreme right-hand root of equation (6.6) is less than

$$m_p(A,B) := w_p(A)\delta_p(w_p(A)/\|B\|),$$

where $\delta_p(a)$ is defined by (3.4). That is,

$$m_p(A,B) = \begin{cases} pe\|B\| & \text{if } w_p(A) \le ep\|B\| \\ w_p(A)\,[ln\,(w_p(A)/p\|B\|)]^{-1/2p} & \text{if } w_p(A) > ep\|B\| \end{cases}.$$

Now Theorem 8.6.3 implies

Corollary 8.6.4 *Let conditions (1.2) and (6.6) hold. Then for any $\mu \in \sigma(A+B)$, there is a $\mu_0 \in \sigma(A)$, such that $|\mu - \mu_0| \le m_p(A,B)$. In particular,*

$$r_s(A+B) \le r_s(A) + m_p(A,B),$$

$$r_l(A+B) \ge \max\{r_l(A) - \tilde{\tau}_p(A,B), 0\} \text{ and } \alpha(A+B) \le \alpha(A) + \tilde{\tau}_p(A,B).$$

8.7 Notes

The material of this chapter is based on the papers (Gil', 2002) and (Gil', 2003). About the well-known perturbations results see (Kato, 1966), (Baumgártel, 1985) and references therein.

References

[1] Baumgártel, H. (1985). *Analytic Perturbation Theory for Matrices and Operators.* Operator Theory, Advances and Appl., 52. Birkháuser Verlag, Basel, Boston, Stuttgart.

[2] Gil', M. I. (2002). Invertibility and spectrum localization of non-selfadjoint operators, *Advances in Applied Mathematics*, **28**, 40-58.

[3] Gil', M. I. (2003). Inner bounds for spectra of linear operators, *Proceedings of the American Mathematical Society* , (to appear)

[4] Kato, T. (1966). *Perturbation Theory for Linear Operators*, Springer-Verlag. New York.

[5] Ostrowski, A. M. (1973). *Solution of Equations in Euclidean and Banach spaces.* Academic Press, New York - London.

9. Spectrum Localization of Nonself-adjoint Operators

In the present chapter we consider operators of the form $A = D + V_+ + V_-$, where D is a normal operator and $V_\pm$ are Volterra (quasinilpotent compact) operators. Numerous integral, integro-differential operators and infinite matrices can be represented in such a form. We investigate the invertibility conditions and bounds for the spectra of the mentioned operators.

9.1 Invertibility Conditions

Let $P(t)$ $(a \leq t \leq b; \; -\infty \leq a < b \leq \infty)$ be a maximal resolution of the identity in a separable Hilbert space H (see Section 7.2). Let

$$A = D + V_+ + V_-, \tag{1.1}$$

where D is a normal operator and $V_\pm$ are Volterra operators satisfying the conditions

$$P(t)V_+P(t) = V_+P(t); \; P(t)V_-P(t) = P(t)V_-,$$

$$P(t)Dh = DP(t)h \; (a \leq t \leq b; \; h \in Dom\,(D)). \tag{1.2}$$

As above, Y is a norm ideal of compact operators in H, which is complete in an auxiliary norm $|\cdot|_Y$. In addition, there are positive numbers θ_k $(k \in \mathbf{N})$ with $\theta_k^{1/k} \to 0$ $(k \to \infty)$, for which, for arbitrary Volterra operators $V \in Y$, $\|V^k\| \leq \theta_k |V|_Y^k$ $(k = 1, 2, ...)$. It is assumed that $V_\pm \in Y$. That is,

$$\|V_\pm^k\| \leq \theta_k |V_\pm|_Y^k \; (k = 1, 2, ...). \tag{1.3}$$

In addition, in this section it is assumed that D is boundedly invertible:

$$r_l(D) = inf \, |\sigma(D)| > 0. \tag{1.4}$$

Since D is normal and $V_\pm$ are Volterra operators, due to Lemma 7.3.3 conditions (1.2) are enough to guarantee that

$$W_\pm := D^{-1}V_\pm$$

are also Volterra operators. Under (1.2)-(1.4), put

$$J_Y(W_\pm) \equiv \sum_{k=0}^{ni(W_\pm)-1} \theta_k |W_\pm|_Y^k, \tag{1.5}$$

where $ni(V)$ again denotes the "nilpotency index" of a quasinilpotent operator V (see Section 7.5). With this notation we have

Theorem 9.1.1 *Under conditions (1.1)-(1.4), let*

$$\zeta_Y(A) := max \, \{\frac{1}{J_Y(W_-)} - \|W_+\|, \frac{1}{J_Y(W_+)} - \|W_-\|\} > 0. \tag{1.6}$$

Then A is boundedly invertible, and

$$\|A^{-1}\| \le \frac{1}{r_l(D)\zeta_Y(A)} \, . \tag{1.7}$$

The proof of Theorem 9.1.1 is presented in the next section. Note that in Theorem 9.1.1, one can replace $J_Y(W_\pm)$ by

$$I_Y(W_\pm) := \sum_{k=0}^{\infty} \theta_k |W_\pm|_Y^k.$$

Consider operators, whose off-diagonal parts belong to the Neumann-Schatten ideal C_{2p} with some integer $p \ge 1$:

$$V_\pm \in C_{2p}. \tag{1.8}$$

Again put

$$\theta_j^{(p)} = \frac{1}{\sqrt{[j/p]!}},$$

where $[x]$ means the integer part of a real number x. For a Volterra operator $V \in C_{2p}$ denote

$$J_p(V) = \sum_{k=0}^{ni(V)-1} \theta_k^{(p)} N_{2p}^k(V).$$

Now Theorem 9.1.1 and Corollary 6.9.4 imply

Corollary 9.1.2 *Let relations (1.2), (1.4) and (1.8) hold. In addition, let*

$$\zeta_{2p}(A) \equiv \max\{\frac{1}{J_p(W_-)} - \|W_+\|, \; \frac{1}{J_p(W_+)} - \|W_-\|\} > 0. \qquad (1.9)$$

Then operator A represented by (1.1) is boundedly invertible. Moreover,

$$\|A^{-1}\| \le \frac{1}{\zeta_{2p}(A)r_l(D)}.$$

It is simple to see that, in this corollary, one can replace $J_p(W_\pm)$ by $I_{2p}(W_\pm)$, where

$$I_{2p}(W_\pm) := \sum_{j=0}^{p-1} \sum_{k=0}^{\infty} \frac{N_{2p}^{j+pk}(W_\pm)}{\sqrt{k!}}.$$

Moreover, put

$$\psi_p(W_\pm) := a_0 \sum_{j=0}^{p-1} N_{2p}^{j}(W_\pm) \; exp \, [b_0 \, N_{2p}^{2p}(W_\pm)].$$

Besides, the constants a_0, b_0 can be taken as in the relations

$$a_0 = \sqrt{\frac{c}{c-1}} \text{ and } b_0 = \frac{c}{2} \text{ for any } c > 1, \text{ in particular, } a_0 = \sqrt{2} \text{ and } b_0 = 1,$$
$$(1.10)$$

or as in the relations

$$a_0 = e^{1/2} \text{ and } b_0 = 1/2. \qquad (1.11)$$

In the next section we also prove

Theorem 9.1.3 *Let relations (1.2), (1.4) and (1.8) hold. In addition, let*

$$\tilde{\zeta}_p(A) := \max\{\frac{1}{\psi_p(W_-)} - \|W_+\|, \; \frac{1}{\psi_p(W_+)} - \|W_-\|\} > 0.$$

Then operator A represented by (1.1) is boundedly invertible. Moreover,

$$\|A^{-1}\| \le \frac{1}{\tilde{\zeta}_p(A)r_l(D)}.$$

9.2   Proofs of Theorems 9.1.1 and 9.1.3

We need the following simple

Lemma 9.2.1 *Under conditions (1.2) and (1.4), let*

$$\theta_0 := \|(D + V_-)^{-1}V_+\| < 1. \qquad (2.1)$$

Then operator A represented by (1.1) is boundedly invertible. Moreover,

$$\|A^{-1}\| \le \frac{\|(D + V_-)^{-1}\|}{1 - \theta_0}. \qquad (2.2)$$

Proof: According to (1.1) we have

$$A = (D + V_-)(I + (D + V_-)^{-1}V_+).\tag{2.3}$$

Thanks to (1.2) and Lemma 7.5.1,

$$\sigma(D + V_\pm) = \sigma(D).\tag{2.4}$$

So according to (1.4), $D + V_\pm$ is invertible. Moreover, under condition (2.1), the operator

$$I + (D + V_-)^{-1}V_+$$

is invertible and

$$\|(I + (D + V_-)^{-1}V_+)^{-1}\| \le \sum_{k=0}^{\infty} \|((D + V_-)^{-1}V_+)^k\| \le$$

$$\sum_{k=0}^{\infty} \theta_0^k = (1 - \theta_0)^{-1}.$$

So due to (2.3)

$$\|A^{-1}\| \le \|(I + (D + V_-)^{-1}V_+)^{-1}\|\|(D + V_-)^{-1}\|.$$

This proves the required result. $\square$

Proof of Theorem 9.1.1: Since $V_\pm \in Y$,

$$\|(D + V_-)^{-1}V_+\| = \|(I + D^{-1}V_-)^{-1}D^{-1}V_+\| = \|(I + W_-)^{-1}W_+\| \le$$

$$\|W_+\| \sum_{k=0}^{\infty} \|W_-^k\| = \|W_+\| \sum_{k=0}^{ni(W_-)-1} \|W_-^k\| \le$$

$$\|W_+\| \sum_{k=0}^{ni(W_-)-1} \theta_k |W_-|_Y^k = \|W_+\| J_Y(W_-).$$

Hence

$$\theta_0 \le \|W_+\| J_Y(W_-).$$

But condition (1.6) implies that at least one of the following inequalities

$$\|W_+\| J_Y(W_-) < 1\tag{2.5}$$

or

$$\|W_-\| J_Y(W_+) < 1\tag{2.6}$$

are valid. If condition (2.5) holds, then (2.1) is valid. Moreover, since D is a normal operator, $\|D^{-1}\| = r_l(D)^{-1}$. Thus,

$$\|(D + V_-)^{-1}\| = \|(I + W_-)^{-1}D^{-1}\| \leq \|D^{-1}\| \sum_{k=0}^{\infty} \|W_-^k\| =$$

$$r_l(D)^{-1} \sum_{k=0}^{ni(W_-)-1} \|W_-^k\| \leq r_l(D)^{-1} \sum_{k=0}^{ni(W_-)-1} \theta_k |W_-|_Y^k = r_l(D)^{-1} J_Y(W_-).$$

Thus, under (2.5), Lemma 9.2.1 yields the inequality

$$\|A_0^{-1}\| \leq \frac{J_Y(W_-)}{r_l(D)(1 - \|W_+\|J_Y(W_-))} = \frac{1}{r_l(D)(J_Y^{-1}(W_-) - \|W_+\|)}. \quad (2.7)$$

Interchanging W_- and W_+, under condition (2.6), we get

$$\|A_0^{-1}\| \leq \frac{1}{r_l(D)(J_Y^{-1}(W_+) - \|W_-\|)}.$$

This relation and (2.7) yield the required result. $\square$

Proof of Theorem 9.1.3: Due to Theorem 6.7.3,

$$\|(D + V_\pm)^{-1}\| = \|(I + W_\pm)^{-1}D^{-1}\| \leq \psi_p(W_\pm)\rho_l^{-1}(D).$$

In addition,

$$\|(D + V_-)^{-1}V_+\| = \|(I + W_-)^{-1}W_+\| \leq \psi_p(W_-)\|W_+\|$$

and

$$\|(D + V_+)^{-1}V_-\| \leq \psi_p(W_+)\|W_-\|.$$

Now the required result is due to Lemma 9.2.1. $\square$

9.3   Resolvents of Quasinormal Operators

For a $V \in Y$, denote

$$J_Y(V, m, z) := \sum_{k=0}^{m-1} z^{-1-k}\theta_k |V|_Y^k \ (z > 0).$$

Due to Lemma 7.3.4, $(D - \lambda I)^{-1}V_\pm$ is a quasinilpotent operator for any $\lambda \notin \sigma(D)$. Put

$$\nu_\pm(\lambda) \equiv ni((D - \lambda I)^{-1}V_\pm).$$

Everywhere below we can replace $\nu_\pm(\lambda)$ by ∞.

Again, $R_\lambda(A)$ is the resolvent and $\rho(D, \lambda) = \inf_{z \in \sigma(D)} |s - z|$.

Lemma 9.3.1 *Under conditions (1.2), (1.3), for a $\lambda \notin \sigma(D)$, let*

$$\zeta(A, \lambda) \equiv \max\{\frac{1}{J_Y(V_-, \nu_-(\lambda), \rho(D, \lambda))} - \|V_+\|,$$

$$\frac{1}{J_Y(V_+, \nu_+(\lambda), \rho(D, \lambda))} - \|V_-\|\} > 0. \tag{3.1}$$

Then λ is a regular point of operator A represented by (1.1). Moreover,

$$\|R_\lambda(A)\| \leq \frac{1}{\zeta(A, \lambda)\rho(D, \lambda)}. \tag{3.2}$$

Proof: Since, D is a normal operator, $\|(D - \lambda I)^{-1}\| = \rho^{-1}(\lambda, D)$. Thus,

$$|(D - \lambda I)^{-1}V_\pm|_Y \leq \|(D - \lambda I)^{-1}\||V_\pm|_Y =$$

$$\rho^{-1}(\lambda, D)|V_\pm|_Y.$$

Hence,

$$\|(D - \lambda I)^{-1}V_+\| \sum_{k=0}^{\nu_-(\lambda)-1} \theta_k|(D - \lambda I)^{-1}V_-|_Y^k \leq$$

$$\|V_+\| \sum_{k=0}^{\nu_-(\lambda)-1} \theta_k \rho^{-1-k}(\lambda, D)|V_-|_Y^k = \|V_+\|J_Y(V_-, \nu_-(\lambda), \rho(D, \lambda)).$$

Similarly,

$$\|(D - \lambda I)^{-1}V_-\| \sum_{k=0}^{\nu_+(\lambda)-1} \theta_k|(D - \lambda I)^{-1}V_+|_Y^k \leq \|V_-\|J_Y(V_+, \nu_+(\lambda), \rho(D, \lambda)).$$

Now Theorem 9.1.1 with

$$A - \lambda I = D + V_+ + V_- - \lambda I$$

instead of A, yields the required result. $\square$

Furthermore, Lemma 9.3.1 implies

Corollary 9.3.2 *Under conditions (1.1), (1.2) and (1.3), for any $\mu \in \sigma(A)$, there is a $\mu_0 \in \sigma(D)$, such that, either $\mu = \mu_0$, or both the inequalities*

$$\|V_+\|J_Y(V_-, \nu_-(\mu), |\mu - \mu_0|) \geq 1 \text{ and}$$

$$\|V_-\|J_Y(V_+, \nu_+(\mu), |\mu - \mu_0|) \geq 1 \tag{3.3}$$

are true.

This result is exact in the following sense: if either $V_- = 0$, or (and) $V_+ = 0$, then due to the latter corollary,

$$\sigma(A) = \sigma(D). \tag{3.4}$$

Now let condition (1.8) hold. Put

$$\tilde{J}_p(V_\pm, m, z) = \sum_{k=0}^{m-1} \frac{\theta_k^{(p)} N_{2p}^k(V_\pm)}{z^{k+1}} \quad (z > 0).$$

Theorem 6.7.1 and Lemma 9.3.1 imply

Corollary 9.3.3 *Under conditions (1.2) and (1.8), for a $\lambda \notin \sigma(D)$, let*

$$\zeta_{2p}(\lambda, A) \equiv \max\{\frac{1}{\tilde{J}_p(V_-, \nu_-(\lambda), \rho(D, \lambda))} - \|V_+\|,$$

$$\frac{1}{\tilde{J}_p(V_+, \nu_+(\lambda), \rho(D, \lambda))} - \|V_-\|\} > 0.$$

Then λ is a regular point of operator A, represented by (1.1). Moreover,

$$\|R_\lambda(A)\| \le \frac{1}{\rho(D, \lambda)\zeta_{2p}(\lambda, A)}.$$

Furthermore, thanks to Theorem 9.1.3 with $A - \lambda$ instead of A, we can replace $\tilde{J}_p$ by the function

$$\psi_p(V_\pm, z) = a_0 \sum_{j=0}^{p-1} \frac{N_{2p}^j(V_\pm)}{z^{j+1}} exp\,[\frac{b_0 N_{2p}^{2p}(V_\pm)}{z^{2p}}],$$

where a_0, b_0 can be taken from (1.10) or from (1.11). Then we have

Corollary 9.3.4 *Under conditions (1.2) and (1.8), for a $\lambda \notin \sigma(D)$, let*

$$\tilde{\zeta}_{2p}(\lambda, A) \equiv \max\{\frac{1}{\psi_p(V_-, \rho(D, \lambda))} - \|V_+\|,$$

$$\frac{1}{\psi_p(V_+, \rho(D, \lambda))} - \|V_-\|\} > 0.$$

Then λ is a regular point of operator A, represented by (1.1). Moreover,

$$\|R_\lambda(A)\| \le \frac{1}{\rho(D, \lambda)\zeta_{2p}(\lambda, A)}.$$

9.4   Upper Bounds for Spectra

Recall that $sv_D(A)$ denotes the spectral variatin of A with respect to D. Put

$$\tau(A) := \min\{\|V_-\|, \|V_+\|\}, \tag{4.1}$$

$$\tilde{V} := \left\{ \begin{array}{ll} V_+ & \text{if } \|V_+\| \geq \|V_-\| \\ V_- & \text{if } \|V_-\| > \|V_+\| \end{array} \right. \tag{4.2}$$

and

$$\tilde{\nu}_0 = \sup_{\lambda \notin \sigma(D)} ni\,((D - \lambda I)^{-1}\tilde{V}).$$

In the sequel one can replace $\tilde{\nu}_0$ by ∞.

Theorem 9.4.1 *Under conditions (1.1), (1.2), let $\tilde{V} \in Y$ and*

$$\|\tilde{V}^k\| \leq \theta_k |\tilde{V}|_Y^k \quad (k = 1, 2, ...). \tag{4.3}$$

Then the equation

$$\tau(A) J_Y(\tilde{V}, \tilde{\nu}_0, z) = 1 \tag{4.4}$$

has a unique positive root $z_Y(A)$. Moreover, $sv_D(A) \leq z_Y(A)$.

Proof: Due to Corollary 9.3.2,

$$\tau(A) J_Y(\tilde{V}, \tilde{\nu}_0, \rho(D, \mu)) \geq 1$$

for any $\mu \in \sigma(A)$. Comparing this inequality with (4.4), we have $\rho(D, \mu) \leq z_Y(A)$. This inequality proves the theorem. $\square$

Lemma 9.4.2 *Under the conditions (1.1), (1.2) and*

$$\tilde{V} \in C_{2p} \quad (p = 1, 2, ...), \tag{4.5}$$

the equation

$$\tau(A) \sum_{j=0}^{p-1} \frac{N_{2p}^j(\tilde{V})}{z^{j+1}} \exp\left[\frac{1}{2}\left(1 + \frac{N_{2p}^{2p}(\tilde{V})}{z^{2p}}\right)\right] = 1. \tag{4.6}$$

has a unique positive root $z_{2p}(A)$. Moreover,

$$sv_D(A) \leq z_{2p}(A). \tag{4.7}$$

Proof: Due to Corollary 9.3.4,

$$\tau(A) \sum_{j=0}^{p-1} \frac{N_{2p}^j(\tilde{V})}{\rho^{j+1}(D, \mu)} \exp\left[\frac{1}{2}\left(1 + \frac{N_{2p}^{2p}(\tilde{V})}{\rho^{2p}(D, \mu)}\right)\right] \geq 1$$

for any $\mu \in \sigma(A)$. Comparing this inequality with (4.6), we get $\rho(D, \mu) \leq z_{2p}(A)$. This inequality proves the required result. $\square$

To estimate $z_{2p}(A)$, substitute $z = x N_{2p}(\tilde{V})$ in (4.6) and use Lemma 8.3.2. Then $z_{2p}(A) \leq \phi_p(A) = N_{2p}(\tilde{V}) \delta_p(a)$, where $a = N_{2p}(\tilde{V})/\tau(A)$ and

$$\delta_p(a) := \begin{cases} pe/a & \text{if } a \leq pe \\ [ln \, (a/p)]^{-1/2p} & \text{if } a > pe \end{cases} . \tag{4.8}$$

That is,

$$\phi_p(A) := \begin{cases} pe\tau(A) & \text{if } N_{2p}(\tilde{V}) \leq \tau(A)pe \\ N_{2p}(\tilde{V})[ln \, (N_{2p}(\tilde{V})/p\tau(A))]^{-1/2p} & \text{if } N_{2p}(\tilde{V}) > \tau(A)pe \end{cases} . \tag{4.9}$$

Now the previous lemma yields

Corollary 9.4.3 *Under conditions (1.1), (1.2) and (4.5), $sv_D(A) \leq \phi_p(A)$. In particular,*

$$r_s(A) \leq r_s(D) + \phi_p(A),$$

provided D is bounded.

In the case $\tilde{V} \in C_2$ we have

$$\delta_1(a) := \begin{cases} e/a & \text{if } a \leq e \\ [ln \, a]^{-1/2} & \text{if } a > e \end{cases} \tag{4.10}$$

and

$$\phi_1(A) := \begin{cases} e\tau(A) & \text{if } N_2(\tilde{V}) \leq \tau(A)e \\ N_2(\tilde{V}) \, [ln \, (N_2(\tilde{V})/\tau(A))]^{-1/2} & \text{if } N_2(\tilde{V}) > \tau(A)e \end{cases} . \tag{4.11}$$

Now (4.7) implies

$$sv_D(A) \leq z_2(A) \leq \phi_1(A). \tag{4.12}$$

Remark 9.4.4 *Everywhere below $\|V_\pm\|$ can be replaced by their upper bounds, since the right roots of equations (4.4) and (4.6) increase, when the coefficients of these equations increase.*

9.5 Inner Bounds for Spectra

Again, let there be a monotonically increasing continuous scalar-valued function $F(z)$ $(z \geq 0)$ with the properties

$$F(0) = 0, \; F(\infty) = \infty \tag{5.1}$$

such that the inequality

$$\|(\lambda I - A)^{-1}\| \leq F(\rho^{-1}(A, \lambda)) \tag{5.2}$$

holds, where $\rho(A, \lambda)$ is the distance between $\sigma(A)$ and a regular point $\lambda \in \mathbf{C}$ of A. Recall that $\tau(A) := \min\{\|V_-\|, \|V_+\|\}$ and denote by $y(\tau, F)$ the unique positive root of the equation

$$\tau(A)F(1/z) = 1 \ (z > 0). \tag{5.3}$$

Now we are in a position to formulate the main result of the present section.

Theorem 9.5.1 *Let A be defined by (1.1) and conditions (1.2) and (5.2) hold. Then*

$$sv_A(D) \leq y(\tau, F). \tag{5.4}$$

The proof of this theorem is presented below. Recall that

$$r_l(A) := \inf \ |\sigma(A)|, \alpha(A) := \sup \ Re \ \sigma(A).$$

Corollary 9.5.2 *Under the hypothesis of Theorem 9.5.1, the following inequalities are true:*

$$r_s(A) \geq \max\{0, r_s(D) - y(\tau, F)\} \ if \ D \ is \ bounded, \tag{5.5}$$

$$r_l(A) \leq r_l(D) + y(\tau, F) \quad and \tag{5.6}$$

$$\alpha(A) \geq \alpha(D) - y(\tau, F) \ if \ \alpha(D) < \infty. \tag{5.7}$$

Indeed, take μ in such a way that $|\mu| = r_s(D)$. Then due to (5.4), there is $\mu_0 \in \sigma(A)$, such that $|\mu_0| \geq r_s(D) - y(\tau, F)$. Hence, (5.5) follows. Similarly, inequality (5.6) can be proved. Furthermore, take μ in such a way that $Re \ \mu = \alpha(D)$. Due to (1.6) for some $\mu_0 \in \sigma(A)$, $|Re \ \mu_0 - \alpha(D)| \leq y(\tau, F)$. So, either $Re \ \mu_0 \geq \alpha(D)$, or $Re \ \mu_0 \geq \alpha(D) - y(\tau, F)$. Thus, inequality (5.7) is also proved.

Proof of Theorem 9.5.1: Take the operator $B_+ = D + V_+$. Then due to (2.4) $\sigma(B_+) = \sigma(D)$. Due to to Lemma 8.4.2, for any $\mu \in \sigma(D)$, there is $\mu_0 \in \sigma(A)$, such that

$$|\mu_0 - \mu| \leq z_-, \tag{5.8}$$

where z_- is the unique positive root of the equation

$$\|V_-\|F(1/z) = 1 \ (z \geq 0).$$

Now, take $B_- = D + V_-$. Then due to relation (2.4), we get $\sigma(B_-) = \sigma(D)$. Similarly to (5.8), we have that for any $\mu \in \sigma(D)$, there is $\mu_0 \in \sigma(A)$, such that

$$|\mu_0 - \mu| \leq z_+, \tag{5.9}$$

where z_+ is the unique positive root of the equation $\|V_+\|F(1/z) = 1$, since $\|A - B_-\| = \|V_+\|$. Relations (5.8) and (5.9) prove the required result. $\square$

9.6 Bounds for Spectra of Hilbert-Schmidt Operators

Assume that

$$A \in C_2. \tag{6.1}$$

Recall that $g(A)$ is defined in Section 6.4. According to Lemma 6.5.1 everywhere below one can replace $g(A)$ by $\sqrt{2}N_2(A_I)$.

Under (6.1), denote by $\tilde{y}_2(A, \tau)$ the unique non-negative root of the equation

$$\frac{\tau(A)}{z} \, exp \, [\frac{1}{2} + \frac{g^2(A)}{2z^2}] = 1, \tag{6.2}$$

where $\tau(A)$ is defined by (4.1).

Theorem 9.6.1 *Let conditions (1.1), (1.2) and (6.1) hold. Then the relations (4.12) and*

$$sv_A(D) \leq \tilde{y}_2(A, \tau)$$

are valid.

Proof: The required result is due to Theorems 6.4.2 and 9.5.1, and Corollary 9.4.3. □

Substitute $z = g(A)x$ in (6.2) and use Lemma 8.3.2. Then we get

$$\tilde{y}_2(A, \tau) \leq \tilde{\Delta}_2(A) := g(A)\delta_1(\frac{g(A)}{\tau(A)}),$$

where $\delta_1(a)$ is defined by (4.10). That is,

$$\tilde{\Delta}_2(A) := \begin{cases} e\tau(A) & \text{if } g(A) \leq e\tau(A) \\ g(A)[ln \, (g(A)/\tau(A))]^{-1/2} & \text{if } g(A) > e\tau(A) \end{cases}. \tag{6.3}$$

Thus Theorem 9.6.1 and relations (4.12) imply

Corollary 9.6.2 *Let relations (1.1), (1.2) and (6.1) hold. Then*

$$sv_D(A) \leq \phi_1(A) \ and \ sv_A(D) \leq \tilde{\Delta}_2(A).$$

In particular,

$$\max\{0, r_s(D) - \tilde{\Delta}_2(A)\} \leq r_s(A) \leq r_s(D) + \phi_1(A),$$

$$\max\{0, r_l(D) - \phi_1(A)\} \leq r_l(A) \leq r_l(D) + \tilde{\Delta}_2(A)$$

and $\alpha(D) - \tilde{\Delta}_2(A) \leq \alpha(A) \leq \alpha(D) + \phi_1(A)$.

9.7   Von Neumann-Schatten Operators

Assume that
$$A \in C_{2p} \text{ for an integer } p > 1 \tag{7.1}$$

and denote by $y_p(A, \tau)$ the unique non-negative root of the equation

$$\tau(A) \sum_{m=0}^{p-1} \frac{(2N_{2p}(A))^m}{z^{m+1}} \, exp \, [\frac{1}{2} + \frac{(2N_{2p}(A))^{2p}}{2z^{2p}}] = 1, \tag{7.2}$$

where $\tau(A)$ is defined by (4.1).

Theorem 9.7.1 *Let conditions (1.1), (1.2) and (7.1) hold. Then the in-equalities (4.7) and $sv_A(D) \leq y_p(A, \tau)$ are valid.*

Proof: The required result is due to Theorems 6.7.4 and 9.5.1, and Corollary 9.4.4. $\square$

Substitute $z = 2N_{2p}(A)x$ in (7.2) and use Lemma 8.3.2. Then we arrive at the inequality

$$y_p(A, \tau) \leq \Delta_p(A) := N_{2p}(A)\delta_p(\frac{2N_{2p}(A)}{\tau(A)}),$$

where $\delta_p(a)$ is defined by (4.8). That is,

$$\Delta_p(A) := \begin{cases} p \, e \, \tau(A) & \text{if } 2N_{2p(A)} \leq p \, e \, \tau(A) \\ 2N_{2p}(A)[ln \, (2N_{2p}(A)/\tau(A))]^{-1/2p} & \text{if } 2N_{2p}(A) > p \, e \, \tau(A) \end{cases}. \tag{7.3}$$

Thus Theorem 9.7.1 and Corollary 9.4.3 imply

Corollary 9.7.2 *Let relations (1.1), (1.2) and (7.1) hold. Then*

$$sv_D(A) \leq \phi_p(A) \text{ and } sv_A(D) \leq \Delta_p(A).$$

In particular,

$$\max\{0, r_s(D) - \Delta_p(A)\} \leq r_s(A) \leq r_s(D) + \phi_p(A),$$

$$\max\{0, r_l(D) - \phi_p(A)\} \leq r_l(A) \leq r_l(D) + \Delta_p(A)$$

and $\alpha(D) - \Delta_p(A) \leq \alpha(A) \leq \alpha(D) + \phi_p(A)$.

9.8   Operators with Hilbert-Schmidt Hermitian Components

In this section it is assumed that $Dom\ (A) = Dom\ (A^*)$ and A has the Hilbert-Schmidt imaginary component $A_I \equiv (A - A^*)/2i$:

$$N_2^2(A_I) = Trace\ (A_I)^2 < \infty. \qquad (8.1)$$

Recall that $g_I(A)$ is defined in Section 7.7. Everywhere below on can replace $g_I(A)$ by $\sqrt{2}N_2(A_I)$.

Under (8.1), denote by $y_H(A, \tau)$ the unique non-negative root of the equation

$$\frac{\tau(A)}{z}\ exp\ [\frac{1}{2} + \frac{g_I^2(A)}{2z^2}] = 1. \qquad (8.2)$$

Theorem 9.8.1 *Let conditions (1.1), (1.2) and (8.1) hold. Then the relations (4.12) and*

$$sv_A(D) \le y_H(A, \tau)$$

are valid.

Proof: The required result is due to Theorems 7.7.1 and 9.5.1, and Corollary 9.4.3. $\square$

Substitute $z = g_I(A)x$ in (8.2) and apply Lemma 8.3.2. Then

$$y_H(A, \tau) \le \Delta_H(A) := g_I(A)\delta_1(\frac{g_I(A)}{\tau(A)}),$$

where $\delta_1(a)$ is defined by (4.10). That is,

$$\Delta_H(A) := \left\{ \begin{array}{ll} e\,\tau(A) & \text{if } g_I(A) \le e\tau(A) \\ g_I(A)[ln\ (g_I(A)/\tau(A))]^{-1/2} & \text{if } g_I(A) > e\,\tau(A) \end{array} \right. .$$

Thanks to (4.12), Theorem 9.8.1 implies

Corollary 9.8.2 *Let relations (1.1), (1.2) and (8.1) hold. Then*

$$sv_D(A) \le \phi_1(A)\ and\ sv_A(D) \le \Delta_H(A).$$

In particular,

$$\max\{0, r_s(D) - \Delta_H(A)\} \le r_s(A) \le r_s(D) + \phi_1(A),$$

$$\max\{0, r_l(D) - \phi_1(A)\} \le r_l(A) \le r_l(D) + \Delta_H(A)$$

and $\alpha(D) - \Delta_H(A) \le \alpha(A) \le \alpha(D) + \phi_1(A)$.

9.9 Operators with Neumann-Schatten Hermitian Components

In this section it is assumed that the Hermitian component $A_I = (A - A^*)/2i$ belongs to the Neumann-Schatten ideal C_{2p} with some integer $p > 1$:

$$N_p(A_I) = [Trace\ A_I^{2p}]^{1/2p} < \infty. \tag{9.1}$$

Recall that $\tilde{\beta}_p$ is defined in Section 7.9. Set

$$w_p(A) := \tilde{\beta}_p N_{2p}(A_I).$$

Let $x_p(A, \tau)$ be the unique positive root of the equation

$$\tau(A) \sum_{m=0}^{p-1} \frac{w_p^m(A)}{z^{m+1}} exp\ [\frac{1}{2} + \frac{w_p^{2p}(A)}{2z^{2p}}] = 1, \tag{9.2}$$

where $\tau(A)$ is defined by (4.1).

Theorem 9.9.1 *Let conditions (1.1), (1.2) and (9.1) hold. Then the relations (4.7) and*

$$sv_A(D) \le x_p(A, \tau)$$

are valid.

Proof: The required result is due to Theorems 7.9.1 and 9.5.1, and Corollary 9.4.3. $\square$

Substitute $z = w_p(A)x$ in (9.2) and apply Lemma 8.3.2. Then

$$x_p(A, \tau) \le m_p(A) := w_p(A)\delta_p(\frac{w_p(A)}{\tau(A)}), \tag{9.3}$$

where $\delta_p(a)$ is defined by (4.8). That is,

$$m_p(A) := \begin{cases} e\tau(A) & \text{if } w_p(A) \le pe\tau(A) \\ w_p(A)[ln\ (w_p(A)/p\tau(A))]^{-1/2p} & \text{if } w_p(A) > pe\tau(A) \end{cases} . \tag{9.4}$$

Thus Theorem 9.9.1 and Corollary 9.4.3 imply

Corollary 9.9.2 *Let relations (1.1), (1.2) and (9.1) hold. Then*

$$sv_D(A) \le \phi_p(A) \text{ and } sv_A(D) \le m_p(A).$$

In particular,

$$\max\{0, r_s(D) - m_p(A)\} \le r_s(A) \le r_s(D) + \phi_p(A),$$

$$\max\{0, r_l(D) - \phi_p(A)\} \le r_l(A) \le r_l(D) + m_p(A)$$

and $\alpha(D) - m_p(A) \le \alpha(A) \le \alpha(D) + \phi_p(A)$.

9.10   Notes

The present chapter is based on the papers (Gil', 2002) and (Gil', 2003). Theorem 9.1.1 supplements the well-known results on the invertibility of linear operators, cf. (Harte, 1988).

As it was above mentioned, a lot of papers and books have been devoted to the spectrum of linear operators. Mainly, the asymptotic distributions of the eigenvalues are considered, cf. the books (Pietsch, 1987), (König, 1986), and references therein. But the bounds and invertibility conditions have been investigated considerably less than the asymptotic distributions. At the same time, in particular, Theorems 9.6.1, 9.7.1 and 9.8.1 and their corollaries give us explicit bounds for the spectrum of the considered operators.

References

[1] Gil', M.I. (2002). Invertibility and spectrum localization of nonselfadjoint operators, *Adv. Appl. Mathematics*, **28**, 40-58.

[2] Gil', M. I. (2003). Inner bounds for spectra of linear operators, *Proceedings of the American Mathematical Society* (to appear).

[3] Harte R. (1988). *Invertibility and Singularity for Bounded Linear Operators*. Marcel Dekker, Inc. New York.

[4] König, H. (1986). *Eigenvalue Distribution of Compact Operators*, Birkhäuser Verlag, Basel- Boston-Stuttgart.

[5] Pietsch, A. (1987). *Eigenvalues and s-Numbers*, Cambridge University Press, Cambridge.

10. Multiplicative Representations of Resolvents

In the present chapter we introduce the notion of the multiplicative operator integral in a separable Hilbert space H. By virtue of the multiplicative operator integral, we derive spectral representations for resolvents of various classes of P-triangular operators. These representations are generalizations of the classical spectral representation for the resolvent of a normal operator. If the maximal resolution of the identity is discrete, then the multiplicative integral is an operator product.

10.1 Operators with Finite Chains of Invariant Projectors

Recall that I is the unit operator in H.

Lemma 10.1.1 *Let P be a projector onto an invariant subspace of a bounded linear operator A in H, $P \neq 0$ and $P \neq I$. Then*

$$\lambda R_\lambda(A) = -(I - APR_\lambda(A)P)(I - A(I - P)R_\lambda(A)(I - P)) \quad (\lambda \notin \sigma(A)).$$

Proof: Denote $E = I - P$. Since

$$A = (E + P)A(E + P) \text{ and } EAP = 0,$$

we have

$$A = PAE + PAP + EAE. \tag{1.1}$$

Let us check the equality

$$R_\lambda(A) = PR_\lambda(A)P - PR_\lambda(A)PAER_\lambda(A)E + ER_\lambda(A)E. \qquad (1.2)$$

In fact, multiplying this equality from the left by $A - I\lambda$ and taking into account the equalities (1.1), $AP = PAP$ and $PE = 0$, we obtain the relation

$$((A - I\lambda)P + (A - I\lambda)E + PAE)(PR_\lambda(A)P-$$

$$PR_\lambda(A)PAER_\lambda(A)E + ER_\lambda(A)E) =$$

$$P - PAER_\lambda(A)E + E + PAER_\lambda(A)E = I.$$

Similarly, multiplying (1.2) by $A - I\lambda$ from the right and taking into account (1.1), we obtain I. Therefore, (1.2) is correct. Due to (1.2)

$$I - AR_\lambda(A) = (I - AR_\lambda(A)P)(I - AER_\lambda(A)E). \qquad (1.3)$$

But

$$I - AR_\lambda(A) = -\lambda R_\lambda(A). \qquad (1.4)$$

We thus arrow at the result. $\square$

Let P_k $(k = 1, \ldots, n)$ be a chain of projectors onto the invariant subspaces of a bounded linear operator A. That is,

$$P_k A P_k = A P_k \quad (k = 1, ..., n) \qquad (1.5)$$

and

$$0 = P_0 H \subset P_1 H \subset ... \subset P_{n-1} H \subset P_n H = H. \qquad (1.6)$$

For bounded linear operators $X_1, X_2, ..., X_n$ again put

$$\overrightarrow{\prod_{1 \le k \le n}} X_k := X_1 X_2 ... X_n.$$

I.e. the arrow over the symbol of the product means that the indexes of the co-factors increase from left to right.

Lemma 10.1.2 *Let a bounded linear operator A have properties (1.5) and (1.6). Then*

$$\lambda R_\lambda(A) = - \overrightarrow{\prod_{1 \le k \le n}} (I - A\Delta P_k R_\lambda(A)\Delta P_k) \quad (\lambda \notin \sigma(A)),$$

where $\Delta P_k = P_k - P_{k-1}$ $(1 \le k \le n)$.

Proof: Due to the previous lemma

$$\lambda R_\lambda(A) = -(I - AP_{n-1}R_\lambda(A)P_{n-1})(I - A(I - P_{n-1})R_\lambda(A)(I - P_{n-1})).$$

But $I - P_{n-1} = \Delta P_n$. So equality (1.4) implies.

$$I - AR_\lambda(A) = (I - AP_{n-1}R_\lambda(A)P_{n-1})(I - A\Delta P_n R_\lambda(A)\Delta P_n).$$

Applying this relation to AP_{n-1}, we get

$$I - AP_{n-1}R_\lambda(A)P_{n-1} =$$

$$(I - AP_{n-2}R_\lambda(A)P_{n-2})(I - A\Delta P_{n-1}R_\lambda(A)\Delta P_{n-1}).$$

Consequently,

$$I - AR_\lambda(A) = (I - AP_{n-2}R_\lambda(A)P_{n-2})(I - A\Delta P_{n-1}R_\lambda(A)\Delta P_{n-1})(I-$$

$$A\Delta P_n R_\lambda(A)\Delta P_n).$$

Continuing this process, we arrive at the required result. $\square$.

Let us consider an operator of the form

$$A = \sum_{k=1}^{n} a_k \Delta P_k + V \tag{1.7}$$

where a_k are some numbers, $\{P_k\}$ is a chain of projectors defined by (1.6) and V is a nilpotent operator with the property

$$P_{k-1}VP_k = VP_k \quad (k = 1, ..., n). \tag{1.8}$$

Lemma 10.1.3 *Under conditions (1.7) and (1.8), the relation*

$$\lambda R_\lambda(A) = - \overrightarrow{\prod_{1 \le k \le n}} \left(I + \frac{A\Delta P_k}{\lambda - a_k}\right)$$

is valid for any $\lambda \ne a_k$ $(k = 1, ..., n)$.

Proof: It is not hard to check that

$$\Delta P_k R_\lambda(A)\Delta P_k = \frac{\Delta P_k}{a_k - \lambda}.$$

Now the required result is due to the previous lemma. $\square$

10.2   Complete Compact Operators

Let A be a compact operator in H whose system of all the root vectors is complete in H. Then there is an orthogonal normed basis (Schur's basis) $\{e_k\}$, such that

$$Ae_k = \sum_{j=1}^{k} a_{jk}e_j, \qquad (2.1)$$

cf. (Gohberg and Krein, 1969, Chapter 5). Moreover $a_{kk} = \lambda_k(A)$ are the eigenvalues of A with their multiplicities. Introduce the orthogonal projectors

$$P_k = \sum_{j=1}^{k}(.,e_j)e_j \ (k=1,2,...).$$

If there exists a limit in the operator norm of the products

$$\prod_{1\leq k\leq n}^{\rightarrow} (I + X_k) \equiv (I + X_1)(I + X_2)...(I + X_n).$$

as $n \to \infty$, then we denote this limit by

$$\prod_{1\leq k\leq \infty}^{\rightarrow} (I + X_k).$$

That is,

$$\prod_{1\leq k\leq \infty}^{\rightarrow} (I + \frac{A\Delta P_k}{\lambda - \lambda_k(A)})$$

is a limit in the operator norm of the sequence the operators

$$\Pi_n(\lambda) := \prod_{1\leq k\leq n}^{\rightarrow} (I + \frac{A\Delta P_k}{\lambda - \lambda_k(A)}) :=$$

$$(I + \frac{A\Delta P_1}{\lambda - \lambda_1(A)})(I + \frac{A\Delta P_2}{\lambda - \lambda_2(A)})...(I + \frac{A\Delta P_n}{\lambda - \lambda_n(A)})$$

for $\lambda \neq \lambda_k(A)$. Here $\Delta P_k = P_k - P_{k-1}, k = 1, 2, ... ; P_0 = 0$, again.

Lemma 10.2.1 *Suppose that the system of all the root vectors of a compact linear operator A is complete in H. Then*

$$\lambda R_\lambda(A) = - \prod_{1\leq k\leq \infty}^{\rightarrow} (I + \frac{A\Delta P_k}{\lambda - \lambda_k(A)}) \ (\lambda \notin \sigma(A)). \qquad (2.2)$$

Proof: Let $A_n = AP_n$. Lemma 10.1.3 implies the equality

$$\lambda R_\lambda(A_n) = -\Pi_n(\lambda). \tag{2.3}$$

Since A is compact, A_n tends to A in the operator norm as n tends to ∞. Besides,

$$(A_n - \lambda I)^{-1} P_n \to (A - \lambda I)^{-1}$$

in the operator norm for any regular λ. We arrive at the result. $\square$

Let A be a normal compact operator. Then

$$A = \sum_{k=1}^{\infty} \lambda_k(A) \Delta P_k.$$

Hence, $A\Delta P_k = \lambda_k(A)\Delta P_k$. Since $\Delta P_k \Delta P_j = 0$ for $j \neq k$, Lemma 10.2.1 gives us the equality

$$-\lambda R_\lambda(A) = I + \sum_{k=1}^{\infty}\left(I + \frac{A\Delta P_k}{\lambda - \lambda_k(A)}\right).$$

But

$$I = \sum_{k=1}^{\infty} \Delta P_k.$$

Thus,

$$\lambda R_\lambda(A) = -\sum_{k=1}^{\infty}[1 + (\lambda - \lambda_k(A))^{-1}\lambda_k(A)]\Delta P_k =$$

$$-\sum_{k=1}^{\infty} \lambda \Delta P_k (\lambda - \lambda_k(A))^{-1}.$$

Or

$$R_\lambda(A) = \sum_{k=1}^{\infty} \frac{\Delta P_k}{\lambda_k(A) - \lambda}.$$

Thus, *Lemma 10.2.1 generalizes the well-known spectral representation for the resolvent of a normal completely continuous operator.*

Furthermore, according to (2.1), the nilpotent part V of A can be defined as

$$V e_k = \sum_{j=1}^{k-1} a_{jk} e_j. \tag{2.4}$$

Therefore, $P_{k-1}V\Delta P_k = P_{k-1}A\Delta P_k = V\Delta P_k$ and

$$A\Delta P_k = P_k A\Delta P_k = \Delta P_k A\Delta P_k + P_{k-1}A\Delta P_k = \lambda_k(A)\Delta P_k + V\Delta P_k.$$

Now Lemma 10.2.1 implies the relation

$$\lambda R_\lambda(A) = -\overrightarrow{\prod_{1 \leq k \leq \infty}} \left(I + \frac{(\lambda_k(A) + V)\Delta P_k}{\lambda - \lambda_k(A)}\right) \quad (\lambda \notin \sigma(A)). \tag{2.5}$$

10.3   The Second Representation for Resolvents of Complete Compact Operators

Let V be a Volterra operator, defined by (2.4). Then due to (2.5)

$$(I - V)^{-1} = \prod_{2 \leq k \leq \infty}^{\rightarrow} (I + V\Delta P_k). \tag{3.1}$$

Furthermore, according to (2.1) $A = D + V$, where D is defined by $De_k = \lambda_k(A)e_k$. Clearly,

$$(A - \lambda I)^{-1} = (D + V - I\lambda)^{-1} = (D - I\lambda)^{-1}(I + B_\lambda)^{-1}, \tag{3.2}$$

where $B_\lambda = V(D - I\lambda)^{-1}$. Due to Lemma 7.3.4 B_λ is a Volterra operator. Moreover, $P_{k-1}B_\lambda P_k = B_\lambda P_k$. Thus relation (3.1) implies

$$(I + B_\lambda)^{-1} = \prod_{2 \leq k \leq \infty}^{\rightarrow} (I - B_\lambda \Delta P_k).$$

But

$$B_\lambda \Delta P_k = \frac{V\Delta P_k}{\lambda_k - \lambda}.$$

Therefore,

$$(I + B_\lambda)^{-1} = \prod_{2 \leq k \leq \infty}^{\rightarrow} (I + \frac{V\Delta P_k}{\lambda - \lambda_k(A)}).$$

Now (3.2) yields

Theorem 10.3.1 *Suppose that the system of all the root vectors of a compact linear operator A is complete in H. Then*

$$R_\lambda(A) = R_\lambda(D) \prod_{2 \leq k \leq \infty}^{\rightarrow} (I + \frac{V\Delta P_k}{\lambda - \lambda_k(A)}) \ \ (\lambda \notin \sigma(A)),$$

where V is the nilpotent part of A and

$$R_\lambda(D) = \sum_{k=1}^{\infty} \frac{\Delta P_k}{\lambda_k(A) - \lambda}.$$

10.4   Operators with Compact Inverse Ones

Let a linear operator A in H have a compact inverse one A^{-1}. Let the system of the root vectors of A^{-1} (and therefore of A) is complete in H. Then due to (2.1), there is an orthogonal normed basis (Schur's basis) $\{e_k\}$, such that

$$A^{-1}e_k = \sum_{j=1}^{k} b_{jk}e_j \tag{4.1}$$

with entries b_{jk}. The nilpotent part V_0 and diagonal one D_0 of A^{-1} are defined by

$$V_0 e_k = \sum_{j=1}^{k-1} b_{jk}e_j. \tag{4.2}$$

and $D_0 e_k = b_{kk}e_k = \lambda_k(A^{-1})e_k$. As above, put

$$P_k = \sum_{j=1}^{k}(.,e_j)e_j \ (k=1,2,...).$$

Theorem 10.4.1 *Let operator A have the compact inverse one A^{-1}. Let the system of the root vectors of $A^{-1}$is complete in H. Then*

$$\lambda \, R_\lambda(A) = \prod_{1\leq k\leq\infty}^{\rightarrow} (I + \frac{\lambda(1+\lambda_k(A)V_0)\Delta P_k}{\lambda_k(A)-\lambda}) - I \ (\lambda \notin \sigma(A)).$$

The product converges in the operator norm.

Proof: Thanks to Lemma 10.2.1,

$$(A-\lambda I)^{-1} = A^{-1}(I-\lambda A^{-1})^{-1} =$$

$$A^{-1} \prod_{1\leq k\leq\infty}^{\rightarrow} (I + \frac{\lambda A^{-1}\Delta P_k}{1-\lambda_k(A^{-1})\lambda})$$

for any regular λ of A. But $D_0\Delta P_k = \lambda_k(A^{-1})\Delta P_k$. Hence,

$$(A-\lambda I)^{-1} = A^{-1} \prod_{1\leq k\leq\infty}^{\rightarrow} (I + \frac{\lambda(\lambda_k(A^{-1})+V_0)\Delta P_k}{1-\lambda_k(A^{-1})\lambda}).$$

Thus, we have derived the relation

$$(A-\lambda I)^{-1} = A^{-1} \prod_{1\leq k\leq\infty}^{\rightarrow} (I + \frac{\lambda(1+\lambda_k(A)V_0)\Delta P_k}{\lambda_k(A)-\lambda}).$$

Taking into account that $A(A-\lambda I)^{-1} = I + \lambda(A-\lambda I)^{-1}$, we arrive at the required result. $\square$

10.5   Multiplicative Integrals

Let F be a function defined on a finite real segment $[a, b]$ whose values are bounded linear operators in H. We define *the right multiplicative integral as the limit in the uniform operator topology of the sequence of the products*

$$\overrightarrow{\prod_{1 \leq k \leq n}} (1 + \delta F(t_k^{(n)})) := (1 + \delta F(t_1^{(n)}))(I + \delta F(t_2^{(n)}))...(I + \delta F(t_n^{(n)}))$$

as $\max_k |t_k^{(n)} - t_{k-1}^{(n)}|$ tends to zero. Here

$$\delta F(t_k^{(n)}) = F(t_k^{(n)}) - F(t_{k-1}^{(n)}) \text{ for } k = 1, ..., n$$

and $a = t_0^{(n)} < t_1^{(n)} < ... < t_n^{(n)} = b$. *The right multiplicative integral we denote by*

$$\int_{[a,b]}^{\overrightarrow{\quad}} (1 + dF(t)).$$

In particular, let P be an orthogonal resolution of the identity defined on $[a, b]$, ϕ be a function integrable in the Riemann-Stieljes with respect to P, and A be a compact linear operator. Then the right multiplicative integral

$$\int_{[a,b]}^{\overrightarrow{\quad}} (I + \phi(t)AdP(t))$$

is the limit in the uniform operator topology of the sequence of the products

$$\overrightarrow{\prod_{1 \leq k \leq n}} (I + \phi(t_k^{(n)})A\Delta P(t_k^{(n)})) \ \ (\Delta P(t_k^{(n)}) = P(t_k^{(n)}) - P(t_{k-1}^{(n)}))$$

as $\max_k |t_k^{(n)} - t_{k-1}^{(n)}|$ tends to zero.

10.6 Resolvents of Volterra Operators

Lemma 10.6.1 *Let V be a Volterra operator with a m.r.i. $P(t)$ defined on a finite real segment $[a, b]$. Then the sequence of the operators*

$$V_n = \sum_{k=1}^{n} P(t_{k-1}^{(n)}) V \Delta P(t_k^{(n)}) \tag{6.1}$$

tends to V in the uniform operator topology as $max_k |t_k^{(n)} - t_{k-1}^{(n)}|$ tends to zero.

Proof: We have

$$V - V_n = \sum_{k=1}^{n} \Delta P(t_k^{(n)}) V \Delta P(t_k^{(n)}).$$

But thanks to the well known Lemma I.3.1 (Gohberg and Krein, 1970), the sequence $\{\|V - V_n\|\}$ tends to zero as n tends to infinity. This proves the required result. $\square$

Lemma 10.6.2 *Let V be a Volterra operator with a maximal resolution of the identity $P(t)$ defined on a segment $[a,b]$. Then*

$$(I - V)^{-1} = \int_{[a,b]}^{\rightarrow} (I + V dP(t)).$$

Proof: Due to Lemma 10.6.1, V is the limit in the operator norm of the sequence of operators V_n, defined by (6.1) . Due to Lemma 10.1.2,

$$(I - V_n)^{-1} = \prod_{1 \leq k \leq n}^{\rightarrow} (I + V_n \Delta P(t_k^{(n)})).$$

Hence the required result follows. $\square$

10.7 Resolvents of P-Triangular Operators

In this section $[a, b]$ is a finite real segment, again.

Theorem 10.7.1 *Let A be a P-triangular operators with a m.r.i. $P(.)$ defined on $[a, b]$, a (compact) nilpotent part V and the diagonal part*

$$D = \int_a^b \phi(t) dP(t), \tag{7.1}$$

where ϕ is a scalar function integrable in the Riemann-Stieljes sense with respect to $P(.)$. Then

$$R_\lambda(A) = R_\lambda(D) \int_{[a,b]}^{\rightarrow} (I - \frac{V dP(t)}{\phi(t) - \lambda}) \quad (\lambda \notin \sigma(A)). \tag{7.2}$$

Proof: By Lemma 7.3.4 $VR_\lambda(D)$ is a Volterra operator. We invoke Lemma 10.6.3. It asserts that

$$(I + VR_\lambda(D))^{-1} = \int_{[a,b]}^{\rightarrow} (I - VR_\lambda(D)dP(t)). \qquad (7.3)$$

But according to (7.1)

$$R_\lambda(D)dP(t) = \frac{1}{\phi(t) - \lambda} dP(t).$$

Thus,

$$(I + VR_\lambda(D))^{-1} = \int_{[a,b]}^{\rightarrow} (I - \frac{VdP(t)}{\phi(t) - \lambda}).$$

Hence relation (3.2) yields the required result. □
   Furthermore, from (7.2) it follows that

$$R_\lambda(A) = \int_a^b \frac{dP(s)}{\phi(s) - \lambda} \int_{[a,b]}^{\rightarrow} (I - \frac{VdP(t)}{\phi(t) - \lambda})$$

for all regular λ. But $dP(s)VdP(t) = 0$ for $t \leq s$. We thus get

Corollary 10.7.2 *Let the hypothesis of Theorem 10.7.1 hold. Then*

$$R_\lambda(A) = \int_a^b \frac{dP(s)}{\phi(s) - \lambda} \int_{[s,b]}^{\rightarrow} (I - \frac{VdP(t)}{\phi(t) - \lambda}) \quad (\lambda \notin \sigma(A)).$$

Let us suppose that A is a normal operator. Then $V = 0$ and Theorem 10.7.1 yields

$$R_\lambda(A) = \int_a^b \frac{dP(s)}{\phi(s) - \lambda}.$$

Thus, *Theorem 10.7.1 generalizes the classical representation for the resolvent of a normal operator.*

Corollary 10.7.3 *Let the hypothesis of Theorem 10.7.1 hold. Then*

$$R_\lambda(A) = R_\lambda(D) \int_{[a,b]}^{\rightarrow} (I - \frac{2i(P(t)A_I - Im\ \phi(t))dP(t)}{\phi(t) - \lambda}) \quad (\lambda \notin \sigma(A)).$$

Indeed, since $A = D + V$, we have $A_I = V_I + D_I$ with

$$D_I = (D - D^*)/2i \text{ and } V_I = (V - V^*)/2i.$$

But

$$P(t)VdP(t) = VdP(t), dP(t)VdP(t) = 0 \text{ and } P(t)V^*dP(t) = 0.$$

Thus, $VdP(t) = 2iP(t)V_I dP(t)$. Moreover, since $D_I dP(t) = Im\ \phi(t)dP(t)$, we get

$$VdP(t) = 2i[P(t)A_I - Im\ \phi(t)]dP(t).$$

Thus, applying Theorem 10.7.1, we get Corollary 10.7.3. In particular, let A have a purely real spectrum. Then Corollary 10.7.3 implies the representation

$$R_\lambda(A) = \int_a^b \frac{dP(s)}{\phi(s) - \lambda} \int_{[s,b]}^{\rightarrow} (I - \frac{2iA_I dP(t)}{\phi(t) - \lambda})$$

for all regular λ. Let A_R, V_R and D_R are the real components of A, V and D, respectively. Repeating the above arguments, by Theorem 10.7.1, we easily obtain the following result.

Corollary 10.7.4 *Let the hypothesis of Theorem 10.7.1 hold. Then*

$$R_\lambda(A) = R_\lambda(D) \int_{[a,b]}^{\rightarrow} (I - \frac{2(P(t)A_R - Re\ \phi(t))dP(t)}{\phi(t) - \lambda})\ \ (\lambda \notin \sigma(A)).$$

10.8   Notes

The contents of Sections 10.1-10.3 and 10.5-10.8 is based on the papers (Gil', 1973) and (Gil', 1980). The results presented in Sections 10.4 and 10.8 are probably new.

   For more details about the multiplicative integral see (Gohberg and Krein, 1970), (Brodskii, 1971), (Feintuch and Saeks, 1982).

References

[1] Brodskii, M. S. (1971). *Triangular and Jordan Representations of Linear Operators*, Transl. Math. Monogr., v. 32, Amer. Math. Soc. Providence, R. I.

[2] Feintuch, A., Saeks, R. (1982). *System Theory. A Hilbert Space Approach*. Ac. Press, New York.

[3] Gil', M. I. (1973). On the representation of the resolvent of a nonselfadjoint operator by the integral with respect to a spectral function, *Soviet Math. Dokl., 14* : 1214-1217.

[4] Gil', M. I. (1980). On spectral representation for the resolvent of linear operators. *Sibirskij Math. Journal, 21*: 231.

[5] Gohberg, I. C. and Krein, M. G. (1970). *Theory and Applications of Volterra Operators in Hilbert Space*, Trans. Mathem. Monogr., v. 24, Amer. Math. Soc., Providence, R. I.

11. Relatively P-Triangular Operators

This chapter is devoted to operators of the type $A = D + W$, where D is a normal boundedly invertible operator in a separable Hilbert space H, and W has the following property: $V := D^{-1}W$ is a Volterra operator in H. If, in addition, A has a maximal resolutions of the identity, then it is called a relatively P-triangular operator. Below we derive estimates for the resolvents of various relatively P-triangular operators and investigate spectrum perturbations of such operators.

11.1 Definitions and Preliminaries

Let $P(.)$ be a maximal resolution of the identity defined on a real segment $[a, b]$ (see Section 7.2). In the present chapter paper we consider a linear operator A in H of the type

$$A = D + W \tag{1.1}$$

where D *is a normal boundedly invertible, generally unbounded operator* and W is linear operator with the properties

$$H \supseteq Dom\ (W) \supset Dom\ (D) = Dom\ (A).$$

In addition,

$$P(t)WP(t)h = WP(t)h \quad (t \in [a, b],\ h \in Dom\ (W)) \tag{1.2}$$

and

$$DP(t)h = P(t)Dh \quad (t \in [a, b],\ h \in Dom\ (A)). \tag{1.3}$$

Moreover,
$$V := D^{-1}W \text{ is a Volterra operator .} \qquad (1.4)$$

We have
$$P(t)D^{-1}WP(t)h = D^{-1}P(t)WP(t)h = D^{-1}WP(t)h$$

$$(t \in [a,b], \ h \in Dom\ (A)).$$

So
$$P(t)VP(t) = VP(t) \ (t \in [a,b]).$$

Definition 11.1.1 *Let relations (1.1)-(1.4) hold. Then A is said to be a relatively P-triangular operator (RPTO), D is the diagonal part of A and V is the relatively nilpotent part of A.*

Everywhere in the present chapter A denotes a relatively P-triangular operator, D denotes its diagonal part and $V \in Y$ denotes the relatively nilpotent part of A.

Recall that Y is an ideal of linear compact operators in H with a norm $|.|_Y$ and the following property: any Volterra operator $V \in Y$ satisfies the inequalities
$$\|V^k\| \le \theta_k |V|_Y^k \ (k = 1, 2, ...), \qquad (1.5)$$

where constants θ_k are independent of V and $\sqrt[k]{\theta_k} \to 0 \ (k \to \infty)$. Under (1.5) put
$$J_Y(V) = \sum_{k=0}^{ni(V)-1} \theta_k |V|_Y^k \ (\theta_0 = 1),$$

where $ni\ (V)$ is the "nilpotency" index (see Definition 1.4.2). In the sequel one can replace $ni\ (V)$ by ∞.

Let $r_l(D) = \inf |\sigma(D)|$, again. Since D is invertible, $r_l(D) > 0$.

Lemma 11.1.2 *Under conditions (1.1)-(1.5), operator A is boundedly invertible. Moreover, $\|A^{-1}\| \le r_l^{-1}(D)J_Y(V)$ and*

$$\|A^{-1}D\| \le J_Y(V).$$

Proof: Clearly, $A = (D + W) = D(I + V)$. So $A^{-1} = (I + V)^{-1}D^{-1}$. Due to (1.5)
$$\|(I + V)^{-1}\| \le J_Y(V).$$

Since D is normal, $\|D^{-1}\| = r_l^{-1}(D)$. This proves the required results. $\square$

11.2 Resolvents of Relatively P-Triangular Operators

Denote

$$w(\lambda, D) \equiv \inf_{t \in \sigma(D)} |\frac{\lambda}{t} - 1|$$

and

$$\nu(\lambda) := ni\left((D - \lambda I)^{-1}W\right) = ni\left((I - D^{-1}\lambda)^{-1}V\right) \ (\lambda \notin \sigma(D)).$$

Under (1.5) put

$$J_Y(V, m, z) = \sum_{k=0}^{m-1} z^{-1-k}\theta_k|V|_Y^k \ (z > 0).$$

Lemma 11.2.1 *Under conditions (1.1)-(1.5), let $\lambda \notin \sigma(D)$. Then λ is a regular point of operator A. Moreover,*

$$\|R_\lambda(A)\| \leq r_l^{-1}(D)J_Y(V, \nu(\lambda), w(\lambda, D)) \tag{2.1}$$

and

$$\|R_\lambda(A)D\| \leq J_Y(V, \nu(\lambda), w(\lambda, D)). \tag{2.2}$$

Proof: According to (1.1),

$$A - \lambda I = (D - \lambda I)(I + (D - \lambda I)^{-1}W).$$

Consequently

$$R_\lambda(A) = (I + R_\lambda(D)W)^{-1}R_\lambda(D).$$

Taking into account that $(D - \lambda I)^{-1} = (I - D^{-1}\lambda)^{-1}D^{-1}$, we have

$$R_\lambda(A) = (I + (I - D^{-1}\lambda)^{-1}D^{-1}W)^{-1}(I - D^{-1}\lambda)^{-1}D^{-1}. \tag{2.3}$$

But D is normal. Therefore,

$$\|(I - D^{-1}\lambda)^{-1}\| \leq \frac{1}{w(\lambda, D)}.$$

Moreover, due to Lemma 7.3.3, $(I - D^{-1}\lambda)^{-1}V$ is a Volterra operator. So

$$\|(I + (I - D^{-1}\lambda)^{-1}V)^{-1}\| \leq \sum_{k=0}^{\nu(\lambda)-1} \|((I - D^{-1}\lambda)^{-1}V)^k\| \leq$$

$$\sum_{k=0}^{\nu(\lambda)-1} \theta_k|((I - D^{-1}\lambda)^{-1}V)^k|_Y \leq$$

$$\sum_{k=0}^{\nu(\lambda)-1} \theta_k w^{-k}(\lambda, D)|V|_Y^k = w(\lambda, D)J_Y(V, \nu(\lambda), w(\lambda, D)).$$

Thus (2.3) yields

$$\|R_\lambda(A)\| \leq \|(I + (I - D^{-1}\lambda)^{-1}V)^{-1}D^{-1}(I - D^{-1}\lambda)^{-1}\| \leq$$

$$J_Y(V, \nu(\lambda), w(\lambda, D))r_l^{-1}(D).$$

Therefore (2.1) is proved. Moreover, thanks to (2.3)

$$\|R_\lambda(A)D\| \leq \|(I + (I - D^{-1}\lambda)^{-1}V)^{-1}(I - D^{-1}\lambda)^{-1}\| \leq$$

$$J_Y(V, \nu(\lambda), w(\lambda, D)).$$

So inequality (2.2) is also proved. $\square$

11.3   Invertibility of Perturbed RPTO

In the sequel Z is a linear operator in H satisfying the condition

$$m(Z) := \|D^{-1}Z\| < \infty. \tag{3.1}$$

Lemma 11.3.1 *Under conditions (1.1)-(1.5) and (3.1), let*

$$J_Y(V)m(Z) < 1. \tag{3.2}$$

Then the operator $A + Z$ is boundedly invertible. Moreover, the inverse operator satisfies the inequality

$$\|(A + Z)^{-1}\| \leq \frac{J_Y(V)}{\rho_l(D)(1 - J_Y(V)m(Z))}.$$

Proof: Due to Lemma 11.1.2 we have

$$\|A^{-1}Z\| = \|A^{-1}DD^{-1}Z\| \leq J_Y(V)m(Z).$$

But

$$(A + Z)^{-1} - A^{-1} = -A^{-1}Z(A + Z)^{-1} = -A^{-1}DD^{-1}Z(A + Z)^{-1}.$$

Hence,

$$\|(A + Z)^{-1}\| \leq \|A^{-1}\| + \|A^{-1}DD^{-1}Z(A + Z)^{-1}\| \leq$$

$$\|A^{-1}\| + J_Y(V)m(Z)\|(A + Z)^{-1}\|$$

and

$$\|(A + Z)^{-1}\| \leq \|A^{-1}\|(1 - J_Y(V)m(Z))^{-1}.$$

Lemma 11.1.2 yields now the required result. $\square$

11.4 Resolvents of Perturbed RPTO

Recall that $J_Y(V, \nu(\lambda), w(\lambda, D))$ is defined in Section 11.2.

Theorem 11.4.1 *Under conditions (1.1)-(1.5) and (3.1), for a $\lambda \notin \sigma(D)$, let*

$$m(Z)J_Y(V, \nu(\lambda), w(\lambda, D)) < 1. \tag{4.1}$$

Then λ is a regular point of operator $A + Z$. Moreover,

$$\|R_\lambda(A + Z)\| \leq \frac{J_Y(V, \nu(\lambda), w(\lambda, D))}{\rho_l(D)(1 - m(Z)J_Y(V, \nu(\lambda), w(\lambda, D)))}.$$

Proof: Due to Lemma 11.2.1,

$$\|R_\lambda(A)Z\| = \|R_\lambda(A)DD^{-1}Z\| \leq m(Z)J_Y(V, \nu(\lambda), w(\lambda, D)).$$

But

$$R_\lambda(A + Z) - R_\lambda(A) = -R_\lambda(A)ZR_\lambda(A + Z).$$

Hence,

$$\|R_\lambda(A + Z)\| \leq \|R_\lambda(A)\| + \|R_\lambda(A)Z\|\|R_\lambda(A + Z)\| \leq$$

$$\|R_\lambda(A)\| + m(Z)J_Y(V, \nu(\lambda), w(\lambda, D))\|R_\lambda(A + Z)\|.$$

Due to (4.1)

$$\|R_\lambda(A + Z)\| \leq \|R_\lambda(A)\|(1 - m(Z)J_Y(V, \nu(\lambda), w(\lambda, D))^{-1}.$$

Now Lemma 11.2.1 yields the required result. □

Theorem 11.4.1 implies

Corollary 11.4.2 *Under conditions (1.1)-(1.5) and (3.1), for any $\mu \in \sigma(A + Z)$, there is a $\mu_0 \in \sigma(D)$, such that, either $\mu = \mu_0$, or*

$$m(Z)J_Y(V, \nu(\mu), |1 - \mu\mu_0^{-1}|) \geq 1. \tag{4.2}$$

11.5 Relative Spectral Variations

Definition 11.5.1 *Let A and B be linear operators in H. Then the quantity*

$$rsv_A(B) := \sup_{\mu \in \sigma(B)} \inf_{\lambda \in \sigma(A)} |1 - \frac{\mu}{\lambda}|$$

will be called the relative spectral variation of B with respect to A. In addition,

$$rhd(A, B) := \max\{rsv_A(B), rsv_B(A)\}$$

is said to be the relative Hausdorff distance between the spectra of A and B.

Put

$$\tilde{\nu}_0 = \sup_{\lambda \notin \sigma(D)} \nu(\lambda) = \sup_{\lambda \notin \sigma(D)} ni\,((D - \lambda I)^{-1}W).$$

In the sequel one can replace $\tilde{\nu}_0$ by ∞.

Theorem 11.5.2 *Under conditions (1.1)-(1.5) and (3.1), the equation*

$$m(Z)J_Y(V, \tilde{\nu}_0, z) = 1 \tag{5.1}$$

has a unique positive root $z_0(Y, V, Z)$. Moreover,

$$rsv_D(A + Z) \le z_0(Y, V, Z). \tag{5.2}$$

Proof: Comparing equation (5.2) with inequality (4.2), we arrive at the required result. □

Lemma 8.3.1 and Theorem 11.5.2 imply

Corollary 11.5.3 *Under conditions (1.1)-(1.5) and (3.1), we have the inequality*

$$rsv_D(A + Z) \le \delta_Y(A, Z), \tag{5.3}$$

where

$$\delta_Y(A, Z) \equiv 2 \max_{j=1,2,\dots} \sqrt[j]{\theta_{j-1}|V|_Y^{j-1}\|Z\|}.$$

Due to (5.3), for any $\mu \in \sigma(A + Z)$, there is a $\mu_0 \in \sigma(D)$, such that

$$1 \le |\mu\mu_0^{-1}| + \delta_Y(A, Z).$$

Thus, $|\mu| \ge |\mu_0|(1 - \delta_Y(A, Z))$. Let $\mu = r_l(A + Z)$. Then we get

Corollary 11.5.4 *Under conditions (1.1)-(1.5) and (3.1), we have the inequality*

$$r_l(A + Z) \ge r_l(D) \max\{0, \ 1 - \delta_Y(A, Z)\}.$$

11.6   Operators with von Neumann-Schatten Relatively Nilpotent Parts

11.6.1   Invertibility conditions

Let A be a relatively P-triangular operator (RPTO). Throughout this section it is assumed that its relatively nilpotent part

$$V := D^{-1}W \in C_{2p} \tag{6.1}$$

for a natural $p \geq 1$. Again put,

$$J_p(V) = \sum_{k=0}^{ni(V)-1} \theta_k^{(p)} N_{2p}^k(V),$$

where

$$\theta_k^{(p)} = \frac{1}{\sqrt{[k/p]!}}$$

and [.] means the integer part. Due to Corollary 6.9.4,

$$\|V^j\| \leq \theta_j^{(p)} N_{2p}^j(V) \ (j = 1, 2, ...). \tag{6.2}$$

Then Lemma 11.1.2 implies that under (1.1)-(1.4) and (6.1),

$$\|A^{-1}\| \leq r_l^{-1}(D)J_p(V) \text{ and } \|A^{-1}D\| \leq J_p(V). \tag{6.3}$$

Moreover, due to Lemma 11.3.1, we get

Lemma 11.6.1 *Under conditions (1.1)-(1.4), (3.1) and (6.1), let*

$$m(Z)J_p(V) < 1.$$

Then operator $A + Z$ is boundedly invertible. Moreover,

$$\|(A + Z)^{-1}\| \leq \frac{J_p(V)}{(1 - m(Z)J_p(V))r_l(D)}.$$

Under (6.1) for a $z > 0$, put

$$\psi_p(V, z) = a_0 \sum_{j=0}^{p-1} \frac{N_{2p}^j(V)}{z^{j+1}} \exp\left[\frac{b_0 N_{2p}^{2p}(V)}{z^{2p}}\right], \tag{6.4}$$

where constants a_0, b_0 can be taken from the relations

$$a_0 = \sqrt{\frac{c}{c-1}}, b_0 = c/2 \text{ for a } c > 1, \tag{6.5}$$

or

$$a_0 = e^{1/2}, b_0 = 1/2. \tag{6.6}$$

Lemma 11.6.2 *Under conditions (1.1)-(1.4), (3.1) and (6.1), let*

$$m(Z)\psi_p(V, 1) < 1.$$

Then operator $A + Z$ is boundedly invertible. Moreover,

$$\|(A + Z)^{-1}\| \leq \frac{\psi_p(V, 1)}{r_l(D)\,(1 - m(Z)\psi_p(V, 1))}.$$

Proof: Due to Theorem 6.7.3,

$$\|(I - V)^{-1}\| \le \psi_p(V, 1). \tag{6.7}$$

But $A^{-1} = (I + V)^{-1}D^{-1}$. So

$$\|A^{-1}\| \le \rho_l^{-1}(D)\psi_p(V, 1) \text{ and } \|A^{-1}D\| \le \psi_p(V, 1).$$

Now, using the arguments of the proof of Lemma 11.3.1, we arrive at the required result. $\square$

11.6.2   The norm of the resolvent

Recall that $\nu(\lambda)$ is defined in Section 11.2. Relation (6.2) and Lemma 11.2.1, under (6.1) imply

$$\|R_\lambda(A)\| \le r_l^{-1}(D)\tilde{J}_p(V, \nu(\lambda), w(\lambda, D)) \text{ and}$$

$$\|R_\lambda(A)D\| \le \tilde{J}_p(V, \nu(\lambda), w(\lambda, D)) \quad (\lambda \notin \sigma(A)),$$

where

$$\tilde{J}_p(V, m, z) = \sum_{k=0}^{m-1} \frac{\theta_k^{(p)} N_{2p}^k(V)}{z^{k+1}} \quad (z > 0).$$

Now thanks to Theorem 11.4.1, we get

Theorem 11.6.3 *Under conditions (1.1)-(1.4), (3.1) and (6.1), let*

$$m(Z)\tilde{J}_p(V, \nu(\lambda), w(\lambda, D)) < 1.$$

Then λ is a regular point of operator $A + Z$. Moreover,

$$\|R_\lambda(A + Z)\| \le \frac{\tilde{J}_p(V, \nu(\lambda), w(\lambda, D))}{r_l(D) \left(1 - m(Z)\tilde{J}_p(V, \nu(\lambda), w(\lambda, D))\right)}.$$

Put

$$\tilde{V}(\lambda) = (I - D^{-1}\lambda)^{-1}V.$$

Thanks to (2.3)

$$R_\lambda(A)D = (I + \tilde{V}(\lambda))^{-1}(I - D^{-1}\lambda)^{-1}.$$

Taking into account (6.7), we get

$$\|R_\lambda(A)\| \le \psi_p(\tilde{V}(\lambda), 1)w(\lambda, D).$$

Hence, we arrive at the inequalities

$$\|R_\lambda(A)\| \le r_l^{-1}(D)\psi_p(V, w(\lambda, D)) \text{ and } \|R_\lambda(A)D\| \le \psi_p(V, w(\lambda, D)) \tag{6.8}$$

for all $\lambda \notin \sigma(A)$. Now, taking into account (6.8) and repeating the arguments of the proof of Theorem 11.4.1, we arrive at the following result

Corollary 11.6.4 *Under conditions (1.1)-(1.4), (3.1) and (6.1), let*

$$m(Z)\psi_p(V, w(\lambda, D)) < 1.$$

Then λ is a regular point of operator $A + Z$. Moreover,

$$\|R_\lambda(A + Z)\| \leq \frac{\psi_p(V, w(\lambda, D))}{r_l(D) \left(1 - m(Z)\psi_p(V, w(\lambda, D))\right)}.$$

11.6.3   Localization of the spectrum

Corollary 11.6.4 immediately yield.

Corollary 11.6.5 *Under conditions (1.1)-(1.4), (3.1) and (6.1), for any $\mu \in \sigma(A + Z)$, there is a $\mu_0 \in \sigma(D)$, such that, either $\mu = \mu_0$, or*

$$m(Z)\psi_p(V, |1 - \frac{\mu}{\mu_0}|) \geq 1. \tag{6.9}$$

Theorem 11.6.6 *Under conditions (1.1)-(1.4), (3.1) and (6.1), the equation*

$$m(Z) \sum_{k=0}^{p-1} \frac{N_{2p}^k(V)}{z^{k+1}} \; exp \; [(1 + \frac{N_{2p}^{2p}(V)}{z^{2p}})/2] = 1 \tag{6.10}$$

has a unique positive root $z_p(V, Z)$. Moreover, $rsv_D(A + Z) \leq z_p(V, Z)$. In particular, the lower spectral radius of $A + Z$ satisfies the inequality

$$r_l(A + Z) \geq r_l(D) \max\{0, 1 - z_p(V, Z)\}.$$

Proof: Comparing (6.9) and (6.10), we have $w(D, \mu) \leq z_p(V, Z)$, as claimed.
□

Substituting the equality $z = N_{2p}(V)x$, in (6.10) and applying Lemma 8.3.2, we get

$$z_p(V, Z) \leq \Delta_p (V, Z),$$

where

$$\Delta_p(V, Z) := \begin{cases} m(Z)pe & \text{if } N_{2p}(V) \leq m(Z)pe \\ N_{2p}(V)[ln \; (N_{2p}(V)/m(Z)p)]^{-1/2p} & \text{if } N_{2p}(V) > m(Z)pe \end{cases}.$$

Thanks to Theorem 11.6.5 we arrive at the following

Corollary 11.6.7 *Under conditions (1.1)-(1.5), (3.1) and (6.1), $rsv_D(A + Z) \leq \Delta_p (V, Z)$. In particular,*

$$r_l(A + Z) \geq r_l(D) \max\{0, 1 - \Delta_p(V, Z)\}.$$

11.7   Notes

This chapter is based on the papers (Gil', 2001) and (Gil', 2002).

References

[1] Gil', M. I. (2001). On spectral variations under relatively bounded perturbations, *Arch. Math.*, **76**, 458-466.

[2] Gil', M. I. (2002). Spectrum localization of infinite matrices, *Mathematical Physics, Analysis and Geometry*, **4**, 379-394 (2002).

12. Relatively Compact Perturbations of Normal Operators

The present chapter is devoted to linear operators of the type $A = D + W_+ + W_-$, where D is a normal invertible operator and $D^{-1}W_\pm$ are Volterra (compact quasinilpotent) operators. Numerous differential and integro-differential operators are examples of such operators. We derive estimates for the resolvents and bounds for the spectra.

12.1   Invertibility Conditions

Let H be a separable Hilbert space. In the present chapter we consider a linear operator A in H of the type

$$A = D + W_+ + W_-, \tag{1.1}$$

where D is *a normal bounded invertible, generally unbounded operator* and $W_\pm$ are linear operators, such that

$$H \supseteq Dom\,(W_\pm) \supset Dom\,(D) = Dom\,(A).$$

Let $P(t)$ $(-\infty \le t \le \infty)$ be *a maximal resolution of the identity (m.r.i.)* (see Section 7.2). It is assumed that

$$P(t)W_+P(t)h = W_+P(t)h \quad (h \in Dom\,(W_+);\ t \in \mathbf{R}), \tag{1.2a}$$

$$P(t)W_-P(t)h = P(t)W_-h \quad (h \in Dom\,(W_-);\ t \in \mathbf{R}) \tag{1.2b}$$

and

$$P(t)Dh = DP(t)h \ (h \in Dom \ (D); \ t \in \mathbf{R}). \qquad (1.2c)$$

Again we use a normed ideal Y of linear compact operators in H with a norm $|.|_Y$ and having the property: any Volterra operator $V \in Y$ satisfies the inequalities $\|V^k\| \le \theta_k |V|_Y^k \ (k = 1, 2, ...)$ where constants θ_k are independent of V and $\sqrt[k]{\theta_k} \to 0 \ (k \to \infty)$. It is assumed that

$$V_\pm \equiv D^{-1} W_\pm$$

are Volterra operators from ideal Y. That is,

$$\|V_\pm^k\| \equiv \|(D^{-1}W_\pm)^k\| \le \theta_k |V_\pm|_Y^k \ (k = 1, 2, ...). \qquad (1.3)$$

Recall that $ni \ (V)$ means the nilpotency index of a quasinilpotent operator V (see Definition 7.5.2). Put

$$J_Y(V_\pm) = \sum_{k=0}^{ni(V_\pm)-1} \theta_k |V_\pm|_Y^k \ (\theta_0 = 1).$$

Lemma 12.1.1 *Under condition (1.2), (1.3), let*

$$\zeta_0(A) \equiv \max \ \{\frac{1}{J_Y(V_-)} - \|V_+\|, \ \frac{1}{J_Y(V_+)} - \|V_-\|\} > 0. \qquad (1.4)$$

Then operator A represented by (1.1) is boundedly invertible. Moreover, the inverse operator satisfies the inequality

$$\|A^{-1}\| \le \frac{1}{r_l(D)\zeta_0(A)}. \qquad (1.5)$$

Recall that $r_l(.)$ denotes the lower spectral radius.
Proof: Condition (1.4) implies that either

$$\|V_+\| J_Y(V_-) < 1 \qquad (1.6)$$

or

$$\|V_-\| J_Y(V_+) < 1.$$

If (1.6) holds, then replacing in Lemma 11.3.1 A by $D + W_-$ and Z by W_+ we get the invertibility and estimate

$$\|A^{-1}\| \le \frac{J_Y(V_-)}{r_l(D)(1 - \|V_+\| J_Y(V_-))} =$$

$$\frac{1}{r_l(D)(J_Y^{-1}(V_-) - \|V_+\|)}. \qquad (1.7)$$

Exchanging V_- and V_+, we have the estimate

$$\|A^{-1}\| \le \frac{1}{r_l(D)(J_Y^{-1}(V_+) - \|V_-\|)}.$$

This and (1.7) imply the required result. $\square$

12.2   Estimates for Resolvents

Under condition (1.3) denote

$$J_Y(V_\pm, m, z) = \sum_{k=0}^{m-1} z^{-1-k} \theta_k |V_\pm|_Y^k \ (z > 0). \tag{2.1}$$

Lemma 12.2.1 *Under conditions (1.2), the operator* $(D - \lambda I)^{-1} W_\pm$ *is quasinilpotent for any* $\lambda \notin \sigma(D)$.

Proof: Condition (1.2a) implies

$$P(t) V_+ P(t) = P(t) D^{-1} W_+ P(t) = D^{-1} P(t) W_+ P(t) =$$

$$D^{-1} W_+ P(t) = V_+ P(t)$$

and

$$(I - D^{-1}\lambda)^{-1} P(t) = P(t)(I - D^{-1}\lambda)^{-1} \ (t \in (-\infty, \infty)).$$

Thus, due to Lemma 7.3.3, the operator $(I - D^{-1}\lambda)^{-1} V_+$ is a Volterra one. But

$$(I - D^{-1}\lambda)^{-1} V_+ = (D - \lambda I)^{-1} W_+.$$

Similarly, we can prove that $(D - \lambda I)^{-1} W_-$ is a Volterra operator. □
Put

$$\nu_\pm(\lambda) \equiv ni((D - \lambda I)^{-1} W_\pm) = ni((I - \lambda D^{-1})^{-1} V_\pm)$$

and

$$w(\lambda, D) \equiv \inf_{t \in \sigma(D)} |\frac{\lambda}{t} - 1|.$$

Now we are in a position to formulate the main result of the chapter

Theorem 12.2.2 *Under conditions (1.2) and (1.3), for a* $\lambda \notin \sigma(D)$, *let*

$$\zeta(A, \lambda) \equiv \max \{ \frac{1}{J_Y(V_-, \nu_-(\lambda), w(\lambda, D))} - \|V_+\|,$$

$$\frac{1}{J_Y(V_+, \nu_+(\lambda), w(\lambda, D))} - \|V_-\|\} > 0. \tag{2.2}$$

Then λ *is a regular point of operator* A *represented by (1.1). Moreover,*

$$\|R_\lambda(A)\| \leq \frac{1}{\zeta(A, \lambda) \rho_l(D)}. \tag{2.3}$$

Proof: Condition (1.5) means that at least one of the following inequalities holds:

$$J_Y(V_-, \nu_-(\lambda), w(\lambda, D)) \|V_+\| < 1$$

or

$$J_Y(V_+, \nu_+(\lambda), w(\lambda, D)) \|V_-\| < 1. \tag{2.4}$$

If (2.4) holds, then replacing in Lemma 11.4.1 operator A by $D + W_-$ and operator Z by W_+ we get the regularity of λ and estimate

$$\|R_\lambda(A)\| \leq \frac{J_Y(V_+, \nu_+(\lambda), w(\lambda, D))}{\rho_l(D)(1 - \|V_-\| J_Y(V_+, \nu_+(\lambda), w(\lambda, D)))}.$$

Or

$$\|R_\lambda(A)\| \leq \frac{1}{\rho_l(D)(J_Y^{-1}(V_+, \nu_+(\lambda), w(\lambda, D)) - \|V_-\|)}.$$

Exchanging V_+ and V_-, we get

$$\|R_\lambda(A)\| \leq \frac{1}{\rho_l(D)(J_Y^{-1}(V_-, \nu_-(\lambda), w(\lambda, D)) - \|V_+\|)}.$$

These inequalities yield the required result. $\square$

12.3   Bounds for the Spectrum

Recall that $rsv_D(A)$ denotes the relative spectral variation of operator A with respect to D. Again put

$$\tau(A) := \min\{\|V_-\|, \|V_+\|\}, \tag{3.1}$$

$$\tilde{V} := \begin{cases} V_+ & \text{if } \|V_+\| \geq \|V_-\|, \\ V_- & \text{if } \|V_-\| > \|V_+\| \end{cases} \tag{3.2}$$

and

$$\tilde{\nu}_0 = \sup_{\lambda \notin \sigma(D)} ni\, ((D - \lambda I)^{-1}\tilde{V}).$$

In the sequel one can replace $\tilde{\nu}_0$ by ∞.

Theorem 12.3.1 *Under conditions (1.1), (1.2), let $\tilde{V} \in Y$ and*

$$\|\tilde{V}^k\| \leq \theta_k |\tilde{V}|_Y^k \quad (k = 1, 2, ...). \tag{3.3}$$

Then the equation

$$\tau(A)J_Y(\tilde{V}, \tilde{\nu}_0, z) = 1 \tag{3.4}$$

has a unique positive root $z_Y(A)$. Moreover,

$$rsv_D(A) \leq z_Y(A). \tag{3.5}$$

Proof: Due to Theorem 12.2.2,

$$\tau(A)J_Y(\tilde{V}, \tilde{\nu}_0, w(D, \mu)) \geq 1$$

for any $\mu \in \sigma(A)$. Comparing this inequality with (3.4), we have $w(D, \mu) \leq z_Y(A)$. This inequality proves the theorem. $\square$

Theorem 12.3.1 and Lemma 8.3.1 give us the following result

Corollary 12.3.2 *Under conditions (1.1), (1.2) and (3.3), we have*

$$rsv_D(A) \le \Delta_Y(A),$$

where

$$\Delta_Y(A) := 2 \max_{j=1,2,\dots} \sqrt[j]{\theta_{j-1}|\tilde{V}|_Y^{j-1}\tau(A)}.$$

Due to (3.3), for all $\mu \in \sigma(A)$, there is $\mu_0 \in \sigma(D)$, such that

$$1 \le |\mu\mu_0^{-1}| + \Delta_Y(A).$$

Thus $|\mu| \ge |\mu_0|(1 - \Delta_Y(A))$. Now the previous corollary yields

Corollary 12.3.3 *Let conditions (1.1), (1.2) and (3.3) hold. Then*

$$r_l(A) \ge r_l(D) \max\{0, \; 1 - \Delta_Y(A)\}.$$

Remark 12.3.4 *Everywhere below $\|V_\pm\|$ can be replaced by upper bounds.*

12.4 Operators with Relatively von Neumann - Schatten Off-diagonal Parts

12.4.1 Invertibility conditions

In this section it is assumed that $V_\pm = D^{-1}W_\pm$ are quasinilpotent operators belonging to the Neumann-Schatten ideal C_{2p} with some integer $p \ge 1$. That is,

$$N_{2p}(V_\pm) \equiv [Trace \; (V_\pm^* V_\pm)^p]^{1/2p} < \infty. \tag{4.1}$$

Put

$$J_p(V_\pm) = \sum_{k=0}^{ni(V)-1} \theta_k^{(p)} N_{2p}^k(V_\pm),$$

where

$$\theta_k^{(p)} = \frac{1}{\sqrt{[k/p]!}}$$

and [.] means the integer part.

Lemma 12.4.1 *Let relations (1.2) and (4.1) hold. In addition, let*

$$\zeta_{2p}(A) := \max\{J_p^{-1}(V_-) - \|V_+\|, \; J_p^{-1}(V_+) - \|V_-\|\} > 0. \tag{4.2}$$

Then operator A represented by (1.1) is boundedly invertible. Moreover,

$$\|A^{-1}\| \le \frac{1}{\zeta_{2p}(A)r_l(D)}. \tag{4.3}$$

Proof: Due to Corollary 6.9.4,

$$\|V^k\| \leq \theta_k^{(p)} N_{2p}^k(V) \ (k = 1, 2, ...). \tag{4.4}$$

for a quasinilpotent operator $V \in C_{2p}$. Now the required result is due to Lemma 12.1.1. $\square$

Clearly,

$$J_p(V_\pm) \leq I_p(V_\pm) := \sum_{j=0}^{p-1} \sum_{k=1}^{\infty} \frac{N_{2p}^{j+pk}(V_\pm)}{\sqrt{k!}}.$$

According to (4.4) one can replace $J_p(V_\pm)$ in (4.2) by $I_p(V_\pm)$.

Furthermore, for a $z > 0$, put

$$\psi_p(V_\pm, z) = \sum_{j=0}^{p-1} \frac{N_p^j(V_\pm)}{z^{j+1}} \ exp \ [(1 + \frac{N_{2p}^{2p}(V_\pm)}{z^{2p}})/2]. \tag{4.5}$$

Then we have

Corollary 12.4.2 *Let relations (1.2) and (4.1) hold. In addition, let*

$$\tilde{\zeta}_{2p}(A) \equiv \max\{\frac{1}{\psi_p(V_-, 1)} - \|V_+\|, \ \frac{1}{\psi_p(V_+, 1)} - \|V_-\|\} > 0.$$

Then operator A represented by (1.1) is boundedly invertible. Moreover,

$$\|A^{-1}\| \leq \frac{1}{\tilde{\zeta}_{2p}(A) r_l(D)}.$$

Indeed, taking into account Lemma 11.6.2 and repeating the arguments of the proof of the previous lemma, we arrive at the required result.

12.4.2   Estimates for resolvents

Under (4.1) denote,

$$\tilde{J}_p(V_\pm, m, z) = \sum_{k=0}^{m-1} \frac{\theta_k^{(p)} N_{2p}^k(V_\pm)}{z^{k+1}} \ (z > 0).$$

Recall that $\nu_\pm(\lambda) \equiv ni \ ((D - \lambda)^{-1} V_\pm) \leq \infty$. Due to Theorem 12.2.2 and inequality (4.4), we get

Theorem 12.4.3 *Under conditions (1.1), (1.2) and (4.1), for a $\lambda \notin \sigma(D)$, let*

$$\tilde{\zeta}_{2p}(\lambda, A) \equiv \max\{\frac{1}{\tilde{J}_p(V_-, \nu_-(\lambda), w(\lambda, D))} - \|V_+\|,$$

$$\frac{1}{\tilde{J}_p(V_+, \nu_+(\lambda), w(\lambda, D))} - \|V_-\|\} > 0.$$

Then λ is a regular point of operator A, represented by (1.1). Moreover,

$$\|R_\lambda(A)\| \leq \frac{1}{\rho_l(D)\tilde{\zeta}_{2p}(\lambda, A)}.$$

As it was shown in Subsection 11.6.2, $\tilde{J}_p(V, \nu_\pm, z)$ can be replaced by $\psi_p(V, z)$ for an arbitrary Volterra operator $V \in C_{2p}$. Now Theorem 12.4.3 yields

Corollary 12.4.4 *Under conditions (1.1), (1.2) and (4.1), for a $\lambda \notin \sigma(D)$, let*

$$\zeta_{2p}^{(1)}(\lambda, A) \equiv \max\{\frac{1}{\psi_p(V_-, w(\lambda, D))} - \|V_+\|,$$

$$\frac{1}{\psi_p(V_+, w(\lambda, D))} - \|V_-\|\} > 0.$$

Then λ is a regular point of operator A, represented by (1.1). Moreover,

$$\|R_\lambda(A)\| \leq \frac{1}{\rho_l(D)\zeta_{2p}^{(1)}(\lambda, A)}. \tag{4.6}$$

12.4.3   Spectrum localization

Let $\tau(A)$ and $\tilde{V}$ be defined by (3.1) and (3.2), respectively. Theorem 12.3.1 and Corollary 12.4.4 yield

Lemma 12.4.5 *Under the conditions (1.1), (1.2) and*

$$\tilde{V} \in C_{2p}, \tag{4.7}$$

the equation

$$\tau(A) \sum_{j=0}^{p-1} \frac{N_{2p}^j(\tilde{V})}{z^{j+1}} \, exp \, [\frac{1}{2}(1 + \frac{N_{2p}(\tilde{V})}{z^{2p}})] = 1, \tag{4.8}$$

has a unique positive root $z_{2p}(A)$. Moreover,

$$rsv_D(A) \leq z_{2p}(A). \tag{4.9}$$

By virtiue of Lemma 8.3.2, we can assert that $z_{2p}(A) \leq \phi_p(A)$, where

$$\phi_p(A) := \begin{cases} pe\tau(A) & \text{if } N_{2p}(\tilde{V}) \leq \tau(A)pe \\ [ln\,(N_{2p}(\tilde{V})/p\tau(A))]^{-1/2p} & \text{if } N_{2p}(\tilde{V}) > \tau(A)pe \end{cases}. \tag{4.10}$$

Now the previous lemma yields

Corollary 12.4.6 *Under conditions (1.1), (1.2) and (4.7), we have $rsv_D(A) \leq \phi_p(A)$. In particular,*

$$r_l(A) \geq r_l(D) \max\{0, \, 1 - \phi_p(A)\}.$$

12.5   Notes

The present chapter is based on the papers (Gil', 2001) and (Gil', 2002).

References

[1] Gil', M. I. (2001). On spectral variations under relatively bounded perturbations, *Arch. Math.*, **76**, 458-466.

[2] Gil', M. I. (2002). Spectrum localization of infinite matrices, *Mathematical Physics, Analysis and Geometry*, **4**, 379-394 (2002).

13. Infinite Matrices in Hilbert Spaces and Differential Operators

The present chapter is concerned with applications of some results from Chapters 7-12 to integro-differential and differential operators, as well as to infinite matrices in a Hilbert space. In particular, we suggest estimates for the spectral radius of an infinite matrix.

13.1 Matrices with Compact off Diagonals

13.1.1 Upper bounds for the spectrum

Let $\{e_k\}_{k=1}^{\infty}$ be an orthogonal normal basis in a separable Hilbert space H. Let A be a linear operator in H represented by a matrix with the entries

$$a_{jk} = (Ae_k, e_j) \ (j, k = 1, 2, ...), \tag{1.1}$$

where $(.,.)$ is the scalar product in H. Then

$$A = D + V_+ + V_-, \tag{1.2}$$

where V_-, V_+ and D are the upper triangular, lower triangular, and diagonal parts of A, respectively:

$$(V_+e_k, e_j) = a_{jk} \text{ for } j < k, \ (V_+e_k, e_j) = 0 \text{ for all } j > k;$$

$$(V_-e_k, e_j) = a_{jk} \text{ for } j > k, \ (V_-e_k, e_j) = 0 \text{ for all } j < k;$$

$$(De_k, e_k) = a_{kk}, \ (De_k, e_j) = 0 \text{ for } j \neq k \ (j, k = 1, 2, ...). \tag{1.3}$$

Let $\{P_k\}_{k=1}^{\infty}$ be a maximal orthogonal resolution of the identity, where P_k are defined by

$$P_k = \sum_{j=1}^{k} (., e_j)e_j \quad (k = 1, 2, ...). \tag{1.4}$$

Simple calculations show that

$$P_k V_+ P_k = V_+ P_k, \quad P_k V_- P_k = P_k V_- \tag{1.5}$$

and

$$P_k Dh = P_k Dh \quad (h \in Dom \, (D)) \quad (k = 1, 2, ...). \tag{1.6}$$

In addition,

$$Dom \, (D) = \{h = (h_k) \in H : \sum_{k=1}^{\infty} |a_{kk} h_k|^2 < \infty; \; h_k = (h, e_k); \; k = 1, 2, ...\}$$

if D is unbounded. We restrict ourselves by the conditions

$$N_2^2(V_+) = \sum_{j=1}^{\infty} \sum_{k=j+1}^{\infty} |a_{jk}|^2 < \infty, N_2^2(V_-) = \sum_{j=2}^{\infty} \sum_{k=1}^{j-1} |a_{jk}|^2 < \infty. \tag{1.7}$$

That is, V_+, V_- are Hilbert-Schmidt matrices, but the results of Chapters 8 and 9 allow us to investigate considerably more general conditions than (1.7). Without any loss of generality, assume that

$$N_2(V_-) \le N_2(V_+). \tag{1.8}$$

The case $N_2(V_-) \ge N_2(V_+)$ can be considered absolutely similarly. Put

$$\tilde{\phi}_1 = \begin{cases} eN_2(V_-) & \text{if } N_2(V_+) \le eN_2(V_-), \\ N_2(V_+)[ln \, (N_2(V_+)/N_2(V_-))]^{-1/2} & \text{if } N_2(V_+) > eN_2(V_-) \end{cases}. \tag{1.9}$$

Due to Corollary 9.4.3 and Remark 9.4.4 we get

Lemma 13.1.1 *Under conditions (1.7), the spectrum of operator $A = (a_{jk})_{j,k}^{\infty}$ is included in the set*

$$\{z \in \mathbf{C} : |a_{kk} - z| \le \tilde{\phi}_1, \; k = 1, 2, ..., \}. \tag{1.10}$$

In particular, the spectral radius of A satisfies the inequality

$$r_s(A) \le \sup_k |a_{kk}| + \tilde{\phi}_1, \tag{1.11}$$

provided D is bounded: $\|D\| \equiv \sup_k |a_{kk}| < \infty$. In addition,

$$\alpha(A) := \sup Re \, \sigma(A) \le \sup Re \, a_{kk} + \tilde{\phi}_1,$$

provided $\sup Re \, a_{kk} < \infty$. So the considered matrix operator is stable (that is, its spectrum is in the open left half plane), if

$$Re \, a_{kk} + \tilde{\phi}_1 < 0 \quad (k = 1, 2, ...).$$

13.1.2   Lower bounds for the spectrum

Now assume that under (1.7), $D_I = (D - D^*)/2i$ is a Hilbert-Schmidt operator:

$$N_2(D_I) = [\sum_{j=1}^{\infty} |a_{jj} - \bar{a}_{jj}|^2]^{1/2}/2 < \infty. \tag{1.12}$$

Then uder (1.7) $A_I = (A - A^*)/2i$ is a Hilbert-Schmidt operator:

$$N_2(A_I) = [\sum_{j=1}^{\infty}\sum_{k=1}^{\infty} |a_{jk} - \bar{a}_{kj}|^2]^{1/2}/2 < \infty.$$

Recall that

$$g_I(A) = [2N_2^2(A_I) - 2\sum_{k=1}^{\infty} |Im\,\lambda_k(A)|^2]^{1/2}.$$

One can replace $g_I(A)$ by $\sqrt{2}N_2(A_I)$. Put

$$\tilde{\Delta}_H(A) := \begin{cases} eN_2(V_-) & \text{if } g_I(A) \le eN_2(V_-) \\ g_I(A)[\,ln\,(g_I(A)/N_2(V_-))\,]^{-1/2} & \text{if } g_I(A) > eN_2(V_-) \end{cases}. \tag{1.13}$$

Due to Corollary 9.8.2, we get

Lemma 13.1.2 *Under conditions (1.7), (1.12), for the matrix $A = (a_{jk})_{j,k}^{\infty}$, the following relations are true:*

$$r_s(A) \ge \max\{0, \sup_k |a_{kk}| - \tilde{\Delta}_H(A)\}, \tag{1.14}$$

$$r_l(A) \le \inf_k |a_{kk}| + \tilde{\Delta}_H(A) \text{ and } \alpha(A) \ge \sup_k Re\,a_{kk} - \tilde{\Delta}_H(A).$$

So A is unstable, provided, $\sup_k\,Re\,a_{kk} - \tilde{\Delta}_H(A) \ge 0$. Note that, according to Corollary 9.6.2, in the case

$$N_2(A) \equiv [\sum_{j,k=1}^{\infty} |a_{jk}|^2]^{1/2} < \infty,$$

one can replace $g_I(A)$ by

$$g(A) = [N_2^2(A) - \sum_{k=1}^{\infty} |\lambda_k(A)|^2]^{1/2} \le [N_2^2(A) - |Trace\,A^2|]^{1/2}.$$

Example 13.1.3 *In the complex Hilbert space $H = L^2[0,1]$, let us consider an operator A defined by*

$$(Au)(x) = u(x) + \int_0^1 K(x,s)u(s)ds \quad (0 \le x \le 1), \qquad (1.15)$$

where K is a scalar Hilbert-Schmidt kernel. Take the orthogonal normal basis

$$e_k(x) = e^{2\pi i k x} \quad (0 \le x \le 1; \ k = 0, \pm 1, \pm 2, ...). \qquad (1.16)$$

Let

$$K(x,s) = \sum_{j,k=-\infty}^{\infty} b_{jk} e_k(x) e_j(s) \qquad (1.17)$$

be the Fourier expansion of K with the Fourier coefficients b_{jk}. Put

$$a_{jk} = (Ae_k, e_j) = b_{jk} \ (j \ne k), \quad \text{and } a_{jj} = (Ae_j, e_j) = 1 + b_{jj}$$

for $j, k = 0, \pm 1, \pm 2, ...)$. Here $(.,.)$ is the scalar product in $L^2[0,1]$. Assume that $1 + b_{jj} \ne 0$ for any integer j. According to (1.3), under consideration we have

$$N^2(V_+) = \sum_{j=-\infty}^{\infty} \sum_{k=j+1}^{\infty} |b_{jk}|^2 < \infty,$$

and

$$N^2(V_-) = \sum_{j=-\infty}^{\infty} \sum_{k=-\infty}^{j-1} |b_{jk}|^2 < \infty.$$

Now relations (1.11) and (1.14) give us the bounds

$$\max\{0, \sup_{k=0,\pm1,\pm2,...} |1 + b_{kk}| - \tilde{\Delta}_H(A)\} \le r_s(A) \le \sup_{k=0,\pm1,\pm2,...} |1 + b_{kk}| + \tilde{\phi}_1.$$

13.2   Matrices with Relatively Compact Off-diagonals

Under (1.1), assume that

$$\rho_l(D) \equiv \inf_{k=1,2,...} |a_{kk}| > 0 \qquad (2.1)$$

and the operators $D^{-1}V_\pm$ are Hilbert-Schmidt ones: $N_2(D^{-1}V_\pm) = v_\pm$, where

$$v_-^2 = \sum_{j=2}^{\infty} \sum_{k=1}^{j-1} \frac{|a_{jk}|^2}{|a_{jj}|^2} < \infty; \ v_+^2 = \sum_{j=1}^{\infty} \sum_{k=j+1}^{\infty} \frac{|a_{jk}|^2}{|a_{jj}|^2} < \infty. \qquad (2.2)$$

Note that the results of Section 12.4 allow us to investigate considerably more general conditions than (2.2). Without any loss of generality assume that

$$v_- \le v_+. \qquad (2.3)$$

The case $v_- \geq v_+$ can be similarly considered . Put

$$\delta(v_-, v_+) := \begin{cases} ev_- & \text{if } v_+ \leq ev_-, \\ v_+[ln\ (v_+/v_-)]^{-1/2} & \text{if } v_+ > ev_- \end{cases} . \qquad (2.4)$$

Due to Lemma 12.4.5 and Corollary 12.4.6, the spectrum of the matrix $A = (a_{jk})_{j,k}^{\infty}$, under conditions (2.2), (2.3) is included in the set

$$\{z \in \mathbf{C} : |1 - \frac{z}{a_{kk}}| \leq \delta(v_-, v_+),\ k = 1, 2, ..., \}. \qquad (2.5)$$

In particular, the lower spectral radius of A satisfies the inequality

$$r_l(A) \geq \max\{0,\ 1 - \delta(v_-, v_+)\} \inf_k |a_{kk}|. \qquad (2.6)$$

13.3   A Nonselfadjoint Differential Operator

In space $H = L^2[0, 1]$, let us consider an operator A defined by

$$(Au)(x) = -\frac{1}{4}\frac{d^2u(x)}{dx^2} + \frac{w(x)}{2}\frac{du(x)}{dx} + l(x)u(x)$$

$$(0 < x < 1, u \in Dom\ (A)) \qquad (3.1)$$

with the domain

$$Dom\ (A) = \{h \in L^2[0, 1] :\ h'' \in L^2[0, 1],\ h(0) = h(1),\ h'(0) = h'(1)\}. \quad (3.2)$$

Here $w(.), l(.) \in L^2[0, 1]$ are scalar functions. So the periodic boundary conditions

$$u(0) = u(1),\ u'(0) = u'(1) \qquad (3.3)$$

are imposed. With the orthogonal normal basis (1.16), let

$$l = \sum_{k=-\infty}^{\infty} \tilde{l}_k e_k \text{ and } w = \sum_{k=-\infty}^{\infty} \tilde{w}_k e_k\ (\tilde{w}_k = (w, e_k),\ \tilde{l}_k = (l, e_k)) \qquad (3.4)$$

be the Fourier expansions of l and w, respectively. Omitting simple calculations, we have

$$(Ae_k, e_j) = i\pi k\tilde{w}_{j-k} + \tilde{l}_{j-k}\ (k \neq j)$$

and

$$(Ae_k, e_k) = \pi^2 k^2 + i\pi k\tilde{w}_0 + \tilde{l}_0\ (j, k = 0, \pm 1, \pm 2, ...).$$

Here $(.,.)$ is the scalar product in $L^2[0, 1]$. Take $Dom\ (D) = Dom\ (A)$ and rewrite operator A as the matrix $(a_{jk})_{j,k=-\infty}^{\infty}$ with the entries

$$a_{kk} = \pi^2 k^2 + i\pi k\tilde{w}_0 + \tilde{l}_0$$

and

$$a_{jk} = i\pi k\tilde{w}_{j-k} + \tilde{l}_{j-k} \ (j \neq k; \ j, k = 0, \pm 1, \pm 2, ...).$$

Assume that

$$r_l(D) = \inf_k |a_{kk}| > 0.$$

Then $N_2(D^{-1}V_+) = v_\pm$, where

$$v_+^2 = \sum_{k=-\infty}^{\infty} \sum_{j=-\infty}^{k-1} |(\pi^2 k^2 + i\pi k\tilde{w}_0 + \tilde{l}_0)^{-1}(i\pi k\tilde{w}_{j-k} + \tilde{l}_{j-k})|^2 =$$

$$\sum_{k=-\infty}^{\infty} \sum_{m=-\infty}^{-1} |(\pi^2 k^2 + i\pi k\tilde{w}_0 + \tilde{l}_0)^{-1}(i\pi k\tilde{w}_m + \tilde{l}_m)|^2 \leq$$

$$2\sum_{k=-\infty}^{\infty} |\pi^2 k^2 + i\pi k\tilde{w}_0 + \tilde{l}_0|^{-2}\pi|k|^2 \sum_{m=-\infty}^{-1} |\tilde{w}_m|^2 +$$

$$2\sum_{k=-\infty}^{\infty} |(\pi^2 k^2 + i\pi k\tilde{w}_0 + \tilde{l}_0)^{-1}|^2 \sum_{m=-\infty}^{-1} |\tilde{l}_m|^2 < \infty$$

since $w, l \in L^2$. Similarly,

$$v_-^2 = \sum_{k=-\infty}^{\infty} \sum_{j=k+1}^{\infty} |(\pi^2 k^2 + i\pi k\tilde{w}_0 + \tilde{l}_0)^{-1}(i\pi k\tilde{w}_{j-k} + \tilde{l}_{j-k})|^2 < \infty.$$

Acccording to (2.5), the spectrum of the operator A defined by (3.1) is included in the set

$$\{z \in \mathbf{C} : |1 - \frac{z}{\pi^2 k^2 + i\pi k\tilde{w}_0 + \tilde{l}_0}| \leq \delta(v_-, v_+), \ k = 0, \pm 1, \pm 2, ..., \},$$

where $\delta(v_-, v_+)$ is defined by (2.4). In particular, the lower spectral radius of A satisfies the inequality

$$r_l(A) \geq \min_k |\pi^2 k^2 + i\pi k\tilde{w}_0 + \tilde{l}_0| \max\{0, 1 - \delta(v_-, v_+)\}.$$

13.4 Integro-differential Operators

In space $H = L^2[0, 1]$ let us consider the operator

$$(Au)(x) = -\frac{d^2 u(x)}{4dx^2} + w(x)u(x) + \int_0^1 K(x, s)u(s)ds$$

$$(u \in Dom\,(A), \ 0 < x < 1) \tag{4.1}$$

with the domain $Dom\ (A)$ defined by (3.2). So the periodic boundary conditions (3.3) hold. Here K is a Hilbert-Schmidt kernel and $w(.) \in L^2[0,1]$ is a scalar-valued function. Take the orthonormal basis (1.16). Let (1.7) and (3.4) be the Fourier expansions of K and of w, respectively. Obviously, for all $j, k = 0, \pm 1, \pm 2, ...,$

$$a_{jk} = (Ae_j, e_k) = \tilde{w}_{j-k} + b_{jk}\ (j \neq k)\text{ and}$$

$$a_{kk} = (Ae_k, e_k) = \pi^2 k^2 + \tilde{w}_0 + b_{kk}.$$

Assume that

$$r_l(D) = \inf_{k=0,\pm 1,\pm 2,...} |\pi^2 k^2 + \tilde{w}_0 + b_{kk}| > 0. \tag{4.2}$$

Then we have $N_2(D^{-1}V_\pm) = v_\pm$ with

$$v_+^2 = \sum_{k=-\infty}^{\infty} \sum_{j=-\infty}^{k-1} |(\pi^2 k^2 + \tilde{w}_0 + b_{kk})^{-1}(\tilde{w}_{j-k} + b_{jk})|^2 < \infty,$$

and

$$v_-^2 = \sum_{k=-\infty}^{\infty} \sum_{j=k+1}^{\infty} |(\pi^2 k^2 + \tilde{w}_0 + b_{kk})^{-1}(\tilde{w}_{j-k} + b_{jk})|^2 < \infty.$$

According to (2.5), the spectrum of the operator A defined by (4.1) is included in the set

$$\{z \in \mathbf{C} : |1 - \frac{z}{\pi^2 k^2 + \tilde{w}_0 + b_{kk}}| \leq \delta(v_-, v_+),\ k = 0, \pm 1, \pm 2, ..., \},$$

where $\delta(v_-, v_+)$ is defined by (2.4).

13.5 Notes

The results presented in this chapter are based on the paper (Gil', 2001). In particular, inequality (1.11) is sharper than the well-known estimate

$$r_s(A) \leq \sup_j \sum_{k=1}^{\infty} |a_{jk}|, \tag{5.1}$$

cf. (Krasnosel'skij et al, 1989, inequality (16.2)), provided

$$\sup_j \sum_{k=1}^{\infty} |a_{jk}| > \sup_k |a_{kk}| + \tilde{\phi}_1(A).$$

For nonnegative matrices the following estimate is well-known, cf. (Krasnosel'skij et al, 1989 inequality (16.15)):

$$r_s(A) \geq \tilde{r}_\infty(A) \equiv \min_{j=1,...,\infty} \sum_{k=1}^{\infty} a_{jk} \tag{5.2}$$

Our relation (1.14) is sharper than estimate (5.2) in the case $|a_{jk}| = a_{jk}$ $(j, k = 1, 2, ...)$, provided

$$\max_k a_{kk} - \tilde{\Delta}_H(A) > \tilde{r}_\infty(A).$$

That is, (1.11) improves estimate (5.1) and (1.14) improves estimate (5.2) for matrices which are "close" to triangular ones.

The results in Section 13.4 supplement the well-known results on differential operators, cf. (Edmunds and Evans, 1990), (Egorov and Kondratiev, 1996), (Locker, 1999) and references therein.

References

[1] Edmunds, D.E. and Evans V.D. (1990). *Spectral Theory and Differential Operators*. Clarendon Press, Oxford.

[2] Egorov, Y. and Kondratiev, V. (1996). *Spectral Theory of Elliptic Operators*. Birkhäuser Verlag, Basel.

[3] Gil', M.I. (2001). Spectrum localization of infinite matrices, *Mathematical Physics, Analysis and Geometry*, **4**, 379-394

[4] Krasnosel'skii, M.A., Lifshits, J. and A. Sobolev (1989). *Positive Linear Systems. The Method of Positive Operators*, Heldermann Verlag, Berlin.

[5] Locker, J. (1999). *Spectral Theory of Nonself-Adjoint Two Point Differential Operators*. Amer. Math. Soc, Mathematical Surveys and Monographs, Volume 73, R.I.

14. Integral Operators in Space L^2

The present chapter is concerned with integral operators in L^2. In particular, we suggest estimates for the spectral radius of an integral operator.

14.1   Scalar Integral Operators

Consider a scalar integral operator A defined in $H = L^2[0,1]$ by

$$(Au)(x) = a(x)u(x) + \int_0^1 K(x,s)u(s)ds \ (u \in L^2[0,1]; \ x \in [0,1]), \quad (1.1)$$

where $a(.)$ is a real bounded measurable function, K is a real Hilbert-Schmidt kernel. Define the maximal resolution of the identity $P(t) \ (-\epsilon \leq t \leq 1; \ \epsilon > 0)$ by

$$(P(t)u)(x) = \begin{cases} 0 & \text{if } -\epsilon \leq t < x, \\ u(x) & \text{if } x \leq t \leq 1 \end{cases}$$

with $x \in [0,1]$. Then, the conditions (1.1) and (1.2) from Section 9.1 are valid with

$$(Du)(x) = a(x)u(x), \ (V_+u)(x) = \int_x^1 K(x,s)u(s)ds,$$

and

$$(V_-u)(x) = \int_0^x K(x,s)u(s)ds \ (u \in L^2[0,1]; \ x \in [0,1]).$$

So

$$N_2^2(V_+) = \int_0^1 \int_x^1 K^2(x,s) \, ds \, dx$$

and

$$N_2^2(V_-) = \int_0^1 \int_0^x K^2(x,s)ds\,dx.$$

Without any loss of generality, assume that

$$N_2(V_-) \le N_2(V_+). \tag{1.2}$$

The case $N_2(V_-) \ge N_2(V_+)$ can be similarly considered. So according to relations (4.1) and (4.2) from Section 9.4, we have $\tau(A) \le N_2(V_-)$ and $\check{V} = V_+$. Put

$$\tilde{\phi}_1 = \begin{cases} eN_2(V_-) & \text{if } N_2(V_+) \le eN_2(V_-) \\ N_2(V_+)[ln\,(N_2(V_+)/\,N_2(V_-))\,]^{-1/2} & \text{if } N_2(V_+) > eN_2(V_-) \end{cases}. \tag{1.3}$$

Due to Corollary 9.4.3 and Remark 9.4.4, the spectrum of operator A is included in the set

$$\{z \in \mathbf{C} : |a(x) - z| \le \tilde{\phi}_1,\ 0 \le x \le 1\}.$$

Hence, the spectral radius of A satisfies the inequality

$$r_s(A) \le \sup_{x\in[0,1]} |a(x)| + \tilde{\phi}_1.$$

In particular, if $a(x) \equiv 0$, then

$$r_s(A) \le \tilde{\phi}_1(A). \tag{1.4}$$

Let us derive the lower estimates for the spectrum. Clearly,

$$N_2^2(A_I) \equiv N_2^2((A - A^*)/2i) = \int_0^1 \int_0^1 |K(x,s) - K(s,x)|^2 ds\,dx/4.$$

Recall that

$$g_I(A) = [2N_2^2(A_I) - 2\sum_{k=1}^{\infty} |Im\,\lambda_k(A)|^2]^{1/2}$$

and one can replace $g_I(A)$ by $\sqrt{2}N_2(A_I)$. Put

$$\tilde{\Delta}_H := \begin{cases} eN_2(V_-) & \text{if } g_I(A) \le eN_2(V_-) \\ g_I(A)[ln\,(g_I(A)/N_2(V_-))\,]^{-1/2} & \text{if } g_I(A) > eN_2(V_-) \end{cases}. \tag{1.5}$$

Due to Corollary 9.8.2, for the integral operator defined by (1.1), the following relations are true:

$$r_s(A) \ge \max\{0,\ \sup_{x\in[0,1]} |a(x)| - \tilde{\Delta}_H\}, \tag{1.6}$$

$$r_l(A) \le \inf_x |a(x)| + \tilde{\Delta}_H \text{ and } \alpha(A) \ge \sup_{x\in[0,1]} Re\,a(x) - \tilde{\Delta}_H.$$

14.2   Matrix Integral Operators with Relatively Small Kernels

Let $\omega \subseteq \mathbf{R}^m$ be a set with a finite Lebesgue measure, and $H \equiv L^2(\omega, \mathbf{C}^n)$ be a Hilbert space of functions defined on ω with values in $\mathbf{C}^n$ and equipped with the scalar product

$$(f, h)_H = \int_\omega (f(s), h(s))_{C^n} ds,$$

where $(., .)_{C^n}$ is the scalar product in $\mathbf{C}^n$. Consider in $L^2(\omega, \mathbf{C}^n)$ the operator

$$(Ah)(x) = Q(x)h(x) + \int_\omega K(x, s)h(s)ds \ (h \in L^2(\omega, \mathbf{C}^n)), \qquad (2.1)$$

where $Q(x)$, $K(x, s)$ are matrix-valued functions defined on ω and $\omega \times \omega$, respectively. It is assumed that Q is bounded measurable and K is a Hilbert-Schmidt kernel. So

$$A = \tilde{Q} + \tilde{K},$$

where

$$(\tilde{Q}h)(x) = Q(x)h(x)$$

and

$$(\tilde{K}h)(x) = \int_\omega K(x, s)h(s)ds \ (x \in \omega).$$

Besides,

$$N_2(\tilde{K}) = [\int_\omega \int_\omega \|K(x, s)\|_{C^n}^2 ds]^{1/2},$$

where $\|.\|_{C^n}$ is the Euclidean norm.

Lemma 14.2.1 *The spectrum of operator A defined by (2.1) lies in the set*

$$\{\lambda \in \mathbf{C} : N_2(\tilde{K}) \sup_{x \in \omega} \|(Q(x) - I_{C^n}\lambda)^{-1}\|_{C^n} \geq 1\}.$$

Proof: Since,

$$A - \lambda I = \tilde{Q} + \tilde{K} - \lambda I = (\tilde{Q} - \lambda I)(I + (\tilde{Q} - \lambda I)^{-1}\tilde{K}),$$

if

$$\|(\tilde{Q} - \lambda I)^{-1}\tilde{K}\|_H < 1,$$

then λ is a regular point. So for any $\mu \in \sigma(A)$,

$$1 \leq \|(\tilde{Q} - \mu I)^{-1}\|_H \|\tilde{K}\|_H \leq \|(\tilde{Q} - \mu I)^{-1}\|_H N_2(\tilde{K}).$$

But

$$\|(\tilde{Q} - \mu I)^{-1}\|_H \leq \sup_{x \in \omega} \|(Q(x) - I_{C^n}\mu)^{-1}\|_{C^n}.$$

This proves the lemma. □

Due to Corollary 2.1.2, for a fixed x we have

$$\|(Q(x) - I_{C^n}\lambda)^{-1}\|_{C^n} \leq \sum_{k=0}^{n-1} \frac{g^k(Q(x))}{\sqrt{k!}\rho^{k+1}(Q(x), \lambda)}. \tag{2.2}$$

Now Lemma 14.2.1 yields

Lemma 14.2.2 *Let operator A be defined by (2.1). Then its spectrum lies in the set*

$$\{\lambda \in \mathbf{C} : N_2(\tilde{K}) \sum_{k=0}^{n-1} \frac{g^k(Q(x))}{\sqrt{k!}\rho^{k+1}(Q(x), \lambda)} \geq 1, \ x \in \omega\}.$$

Corollary 14.2.3 *Let operator A be defined by (2.1). In addition, let*

$$N_2(\tilde{K}) \sup_{x \in \omega} \sum_{k=0}^{n-1} \frac{g^k(Q(x))}{\sqrt{k!}d_0^{k+1}(Q(x))} < 1,$$

where

$$d_0(Q(x)) = \min_{k=1,\dots,n} |\lambda_k(Q(x))|. \tag{2.3}$$

Then A is boundedly invertible in $L^2(\omega, \mathbf{C}^n)$.

With a fixed $x \in \omega$, consider the algebraic equation

$$z^n = N_2(\tilde{K}) \sum_{k=0}^{n-1} \frac{g^k(Q(x))z^{n-k-1}}{\sqrt{k!}}. \tag{2.4}$$

Lemma 14.2.4 *Let $z_0(x)$ be the extreme right (unique positive) root of (2.4). Then for any point $\mu \in \sigma(A)$ there are $x \in \omega$ and an eigenvalue $\lambda_j(Q(x))$ of matrix $Q(x)$, such that*

$$|\mu - \lambda_j(Q(x))| \leq z_0(x). \tag{2.5}$$

In particular,

$$r_s(A) \leq \sup_x (r_s(Q(x)) + z_0(x)).$$

Proof: Due to Lemma 14.2.2, for any point $\mu \in \sigma(A)$ there is $x \in \omega$, such that the inequality

$$N_2(\tilde{K}) \sum_{k=0}^{n-1} \frac{g^k(Q(x))}{\sqrt{k!}\rho^{k+1}(Q(x), \mu)} \geq 1$$

is valid. Comparing this with (2.4), we have $\rho(Q(x), \mu) \leq z_0(x)$. This proves the required result. □

Corollary 14.2.5 *Let $Q(x)$ be a normal matrix for all $x \in \omega$. Then for any point $\mu \in \sigma(A)$ there are $x \in \omega$ and $\lambda_j(Q(x))\sigma(Q(x))$, such that*

$$|\mu - \lambda_j(Q(x))| \le N_2(\tilde{K}).$$

In particular, $r_s(A) \le N_2(\tilde{K}) + \sup_x(r_s(Q(x)))$.

Indeed, since $Q(x)$ is normal, we have $g(Q(x)) = 0$ and $z_0(x) = N_2(\tilde{K})$. Now the result is due to the latter theorem.

Put

$$b(x) := N_2(\tilde{K}) \sum_{k=0}^{n-1} \frac{g^k(Q(x))}{\sqrt{k!}}.$$

Due to Corollary 1.6.2, $z_0(x) \le \delta_n(x)$, where

$$\delta_n(x) = \sqrt[n]{b(x)} \text{ if } b(x) \le 1 \text{ and } \delta_n(x) = b(x) \text{ if } b(x) > 1.$$

Now Theorem 14.2.4 implies

Theorem 14.2.6 *Under condition (2.7), for any point $\mu \in \sigma(A)$, there are $x \in \omega$ and an eigenvalue $\lambda_j(Q(x))$ of $Q(x)$, such that*

$$|\mu - \lambda_j(Q(x))| \le \delta_n(x).$$

In particular, $r_s(A) \le \sup_x(r_s(Q(x)) + \delta_n(x))$.

14.3   Perturbations of Matrix Convolutions

Consider in $H = L^2([-\pi, \pi], \mathbf{C}^n)$ the convolution operator

$$(Ch)(x) = Q_0 h(x) + \int_{-\pi}^{\pi} K_0(x - s)h(s)ds \; (h \in L^2([-\pi, \pi], \mathbf{C}^n)), \quad (3.1)$$

where Q_0 is a constant matrix, K_0 is a matrix-valued function defined on $[-\pi, \pi]$ with

$$\|K_0\|_{C^n} \in L^2[-\pi, \pi],$$

having the Fourier expansion

$$K_0(x) = \sum_{k=-\infty}^{\infty} D_k e^{ikx}$$

with the matrix Fourier coefficients

$$D_k = \frac{1}{\sqrt{2\pi}} \int_{-\pi}^{\pi} K_0(s) e^{-iks} ds.$$

Put

$$B_k = Q_0 + D_k.$$

We have

$$Ce^{ikx} = B_k e^{ikx}. \tag{3.2}$$

Let d_{jk} be an eigenvector of B_k, corresponding to an eigenvalue $\lambda_j(B_k)$ ($j = 1, ... n$). Then

$$Ce^{ikx} d_{jk} = e^{ikx} Q_0 d_{jk} + \int_{-\pi}^{\pi} K_0(x-s) d_{jk} e^{iks} ds =$$

$$e^{ikx} B_k d_{jk} = e^{ikx} \lambda_j(B_k) d_{jk}.$$

Since the set

$$\{e^{ikx}\}_{k=-\infty}^{k=\infty}$$

is a basis in $L^2[-\pi, \pi]$ we have the following result

Lemma 14.3.1 *The spectrum of operator (3.1) consists of the points*

$$\lambda_j(B_k) \ (k = 0, \pm 1, \pm 2, ... \ ; j = 1, ... n).$$

Let P_k be orthogonal projectors defined by

$$(P_k h)(x) = e^{ikx} \frac{1}{2\pi} \int_{-\pi}^{\pi} h(s) e^{-iks} ds.$$

Since

$$\sum_{k=-\infty}^{\infty} P_k = I_H,$$

it can be directly checked by (3.2) that the equality

$$C = \sum_{k=-\infty}^{\infty} B_k P_k$$

holds. Hence, the relation

$$(C - I_H \lambda)^{-1} = \sum_{k=-\infty}^{\infty} (B_k - I_{C^n} \lambda)^{-1} P_k$$

is valid for any regular λ. Therefore,

$$\|(C - I_H \lambda)^{-1}\|_H \leq \sup_{k=0, \pm 1, ...} \|(B_k - I_{C^n} \lambda)^{-1}\|_{C^n}.$$

Using Corollary 2.1.2, we get

Lemma 14.3.2 *The resolvent of convolution C defined by (3.1) satisfies the inequality*

$$\|(C - \lambda I)^{-1}\|_H \leq \sup_{l=0, \pm 1, \ldots} \sum_{k=0}^{n-1} \frac{g^k(B_l)}{\sqrt{k!}\rho^{k+1}(B_l, \lambda)}.$$

Consider now the operator

$$(Ah)(x) \equiv Q_0 h(x) + \int_{-\pi}^{\pi} K_0(x - s)h(s)ds + (Zh)(x) \ (-\pi \leq x \leq \pi). \quad (3.3)$$

where Z is a bounded operator in $L^2([-\pi, \pi], \mathbf{C}^n)$. We easily have by the previous lemma that the inequalities

$$\|Z\|_H \|(C - \lambda I)^{-1}\|_H \leq \|Z\|_H \sup_{l=0, \pm 1, \ldots} \sum_{k=0}^{n-1} \frac{g^k(B_l)}{\sqrt{k!}\rho^{k+1}(B_l, \lambda)} < 1$$

imply that λ is a regular point. Hence we arrive at

Lemma 14.3.3 *The spectrum of operator A defined by (3.3) lie in the set*

$$\{\lambda \in \mathbf{C} : \|Z\|_H \sup_{l=0, \pm 1, \ldots} \sum_{k=0}^{n-1} \frac{g^k(B_l)}{\sqrt{k!}\rho^{k+1}(B_l, \lambda)} \geq 1\}.$$

In other words, for any $\mu \in \sigma(A)$, there are

$$l = 0, \pm 1, \pm 2, \ldots \ and \ j = 1, \ldots, n,$$

such that

$$\|Z\|_H \sum_{k=0}^{n-1} \frac{g^k(B_l)}{\sqrt{k!}|\mu - \lambda_j(B_l)|^{k+1}} \geq 1.$$

Corollary 14.3.4 *Operator A defined by (3.3) is invertible provided that*

$$\|Z\|_H \sum_{k=0}^{n-1} \frac{g^k(B_k)}{\sqrt{k!}|\lambda_j(B_l)|^{k+1}} \leq c_0 < 1 \ (c_0 = const)$$

for all

$$l = 0, \pm 1, \pm 2, \ldots \ and \ j = 1, \ldots, n.$$

Let z_l be the extreme right (unique positive) root of the equation

$$z^n = \|Z\|_H \sum_{k=0}^{n-1} \frac{z^{n-1-k} g^k(B_l)}{\sqrt{k!}}. \quad (3.4)$$

Since the function in the right part of (3.4) monotonically increases as $z > 0$ increases, Lemma 14.3.4 implies

Theorem 14.3.5 *For any point μ of the spectrum of operator (3.3), there are indexes $l = 0, \pm 1, \pm 2, \ldots$ and $j = 1, \ldots, n$, such that*

$$|\mu - \lambda_j(B_l)| \le z_l, \tag{3.5}$$

where z_l is the extreme right (unique positive) root of the algebraic equation (3.4). In particular,

$$r_s(A) \le \max_{l=0,\pm 1, \ldots} r_s(B_l) + z_l.$$

If all the matrices B_l are normal, then $g(B_l) \equiv 0$, $z_l = \|Z\|_H$, and (3.5) takes the form

$$|\mu - \lambda_j(B_l)| \le \|Z\|_H.$$

Assume that

$$b_l := \|Z\|_H \sum_{k=0}^{n-1} \frac{g^k(B_l)}{\sqrt{k!}} \le 1 \quad (l = 0, \pm 1, \pm 2, \ldots). \tag{3.6}$$

Then due to Lemma 1.6.1

$$z_l \le \sqrt[n]{b_l}.$$

Now Theorem 14.3.5 implies

Corollary 14.3.6 *Let A be defined by (3.3) and condition (3.6) hold. Then for any $\mu \in \sigma(A)$ there are $l = 0, \pm 1, \pm 2, \ldots$ and $j = 1, \ldots, n$, such that*

$$|\mu - \lambda_j(B_l)| \le \sqrt[n]{b_l}.$$

In particular,

$$r_s(A) \le \sup_{l=0,\pm 1, \pm 2, \ldots} \sqrt[n]{b_l} + r_s(B_l).$$

14.4   Notes

Inequality (1.4) improves the well-known estimate

$$r_s(A) \le \tilde{\delta}_0(A) \equiv vrai \sup_x \int_0^1 |K(x,s)| ds,$$

cf. (Krasnosel'skii et al., 1989, Section 16.6) for operators which are "close" to Volterra ones.

     The material in this chapter is taken from the papers (Gil', 2000), (Gil', 2003).

References

[1] Gil', M.I. (2000). Invertibility conditions and bounds for spectra of matrix integral operators, *Monatshefte für mathematik*, **129**, 15-24.

[2] Gil', M.I. (2003). Inner bounds for spectra of linear operators, *Proceedings of the American Mathematical Society* , (to appear).

[3] Krasnosel'skii, M. A., J. Lifshits, and A. Sobolev (1989). *Positive Linear Systems. The Method of Positive Operators*, Heldermann Verlag, Berlin.

[4] Pietsch, A. (1987). *Eigenvalues and s-Numbers*, Cambridge University Press, Cambridge.

15. Operator Matrices

In the present chapter we consider the invertibility and spectrum of matrices, whose entries are unbounded, in general, operators. In particular, under some restrictions, we improve the Gershgorin-type bounds. Applications to matrix differential operators are also discussed.

15.1 Invertibility Conditions

Let H be an orthogonal sum of Hilbert spaces E_k $(k = 1, ..., n < \infty)$ with norms $\|.\|_{E_k}$:
$$H \equiv E_1 \oplus E_2 \oplus ... \oplus E_n.$$

Consider in H the operator matrix

$$A = \begin{pmatrix} A_{11} & A_{12} & ... & A_{1n} \\ A_{21} & A_{22} & ... & A_{2n} \\ . & ... & . & . \\ A_{n1} & A_{n2} & ... & A_{nn} \end{pmatrix}, \tag{1.1}$$

where A_{jk} are linear operators acting from E_k to E_j. In the present chapter, invertibility conditions and bounds for the spectrum of operator (1.1) are investigated under the assumption that we have an information about the spectra of diagonal operators.

Let $h = (h_k \in E_k)_{k=1}^n$ be an element of H. *Everywhere in the present chapter the norm in H is defined by the relation*

$$\|h\| \equiv \|h\|_H = [\sum_{k=1}^n \|h_k\|_{E_k}^2]^{1/2} \tag{1.2}$$

and $I = I_H$ is the unit operator in H.

Denote by V, W and D the upper triangular, lower triangular, and diagonal parts of A, respectively. That is,

$$
V = \begin{pmatrix} 0 & A_{12} & \dots & A_{1n} \\ 0 & 0 & \dots & A_{2n} \\ . & \dots & . & . \\ 0 & 0 & \dots & 0 \end{pmatrix},
$$

$$
W = \begin{pmatrix} 0 & 0 & \dots & 0 & 0 \\ A_{21} & 0 & \dots & 0 & 0 \\ . & \dots & . & . \\ A_{n1} & A_{n2} & \dots & A_{n,n-1} & 0 \end{pmatrix}
$$

and

$$
D = diag\,[A_{11}, A_{22}, ..., A_{nn}].
$$

Recall that, for a linear operator A, $Dom\,(A)$ means the domain, $\sigma(A)$ is the spectrum, $\lambda_k(A)$ $(k = 1, 2, ...)$ are the eigenvalues with their multiplicities, $\rho(A, \lambda)$ is the distance between the spectrum of A and a $\lambda \in \mathbf{C}$.

Theorem 15.1.1 *Let the diagonal operator D be invertible and the operators*

$$
V_A \equiv D^{-1}V, W_A \equiv D^{-1}W \text{ be bounded }. \tag{1.3}
$$

In addition, let the condition

$$
\|\sum_{j,k=1}^{n-1} (-1)^{k+j} V_A^k W_A^j\| < 1 \tag{1.4}
$$

hold. Then operator A defined by (1.1) is invertible.

Proof: We have

$$
A = D + V + W = D(I + V_A + W_A) = D[(I + V_A)(I + W_A) - V_A W_A].
$$

Simple calculations show that

$$
V_A^n = W_A^n = 0. \tag{1.5}
$$

So V_A and W_A are nilpotent operators and, consequently, the operators, $I + V_A$ and $I + W_A$ are invertible. Thus,

$$
A = D(I + V_A)[I - (I + V_A)^{-1}V_A W_A(I + W_A)^{-1}](I + W_A).
$$

Therefore, the condition

$$
\|(I + V_A)^{-1}V_A W_A(I + W_A)^{-1}\| < 1
$$

provides the invertibility of A. But according to (1.5),

$$(I + V_A)^{-1}V_A = \sum_{k=1}^{n-1}(-1)^{k-1}V_A^k, \ W_A(I + W_A)^{-1} = \sum_{k=1}^{n-1}(-1)^{k-1}W_A^k.$$

Hence, the required result follows. $\square$

Corollary 15.1.2 *Let operator matrix A defined by (1.1) be an upper triangular one ($W=0$), D an invertible operator and $D^{-1}V$ a bounded one. Then A is invertible.*

Similarly, let operator matrix A be a lower triangular one ($V=0$) and $D^{-1}W$ a bounded operator. Then A is invertible.

Corollary 15.1.3 *Let the diagonal operator D be invertible and the conditions (1.3) and*

$$\|V_AW_A\| \sum_{j,k=0}^{n-2} \|V_A\|^k \|W_A\|^j < 1 \tag{1.6}$$

hold. Then operator (1.1) is invertible.

In particular, let $\|V_A\|, \|W_A\| \neq 1$. Then (1.6) can be written in the form

$$\|V_AW_A\| \frac{(1 - \|V_A\|^{n-1})(1 - \|W_A\|^{n-1})}{(1 - \|V_A\|)(1 - \|W_A\|)} < 1. \tag{1.7}$$

Indeed, taking into account that

$$\sum_{k=0}^{n-2} \|V_A\|^k = \frac{1 - \|V_A\|^{n-1}}{1 - \|V_A\|}, \sum_{k=0}^{n-2} \|W_A\|^k = \frac{1 - \|W_A\|^{n-1}}{1 - \|W_A\|}$$

and using Theorem 15.1.1, we arrive at the required result. $\square$

We need also the following

Lemma 15.1.4 *Let*

$$a_{jk} \equiv \|A_{jk}\|_{E_k \to E_j} < \infty \ (j, k = 1, ..., n).$$

Then the norm of operator A defined by (1.1) is subject to the relation

$$\|A\| \leq \|\tilde{a}\|_{C^n},$$

where $\tilde{a}$ is the linear operator in the Euclidean space $\mathbf{C}^n$, defined by the matrix with the entries a_{jk} and $\|.\|_{C^n}$ is the Euclidean norm.

The proof is a simple application of relation (1.2) and it is left to the reader.
    The latter corollary implies

$$\|A\|^2 \leq \sum_{j,k=1}^{n} \|A_{jk}\|_{E_k \to E_j}^2. \tag{1.8}$$

Consider the case $n = 2$:

$$A = \begin{pmatrix} A_{11} & A_{12} \\ A_{21} & A_{22} \end{pmatrix}. \tag{1.9}$$

Clearly,

$$V_A = \begin{pmatrix} 0 & A_{11}^{-1}A_{12} \\ 0 & 0 \end{pmatrix} \text{ and } W_A = \begin{pmatrix} 0 & 0 \\ A_{22}^{-1}A_{21} & 0 \end{pmatrix}.$$

Hence,

$$V_A W_A = \begin{pmatrix} A_{11}^{-1}A_{12}A_{22}^{-1}A_{21} & 0 \\ 0 & 0 \end{pmatrix}.$$

Thus, due to Theorem 15.1.1, if

$$\|A_{11}^{-1}A_{12}A_{22}^{-1}A_{21}\| < 1,$$

then operator (1.9) is invertible.

15.2   Bounds for the Spectrum

Theorem 15.2.1 *For any regular point λ of D, let*

$$\tilde{V}(\lambda) := (D - I_H\lambda)^{-1}V \text{ and } \tilde{W}(\lambda) := (D - I_H\lambda)^{-1}W \text{ be bounded operators .} \tag{2.1}$$

Then the spectrum of operator A defined by (1.1) lies in the union of the sets $\sigma(D)$ and

$$\{\lambda \in \mathbf{C} : \|\tilde{V}(\lambda)\tilde{W}(\lambda)\| \sum_{j,k=0}^{n-2} \|\tilde{V}(\lambda)\|^k \|\tilde{W}(\lambda)\|^j \geq 1\}.$$

Indeed, if for some $\lambda \in \sigma(A)$,

$$\|\tilde{V}(\lambda)\tilde{W}(\lambda)\| \sum_{j,k=0}^{n-2} \|\tilde{V}(\lambda)\|^k \|\tilde{W}(\lambda)\|^j < 1, \tag{2.2}$$

then due to Corollary 15.1.3, $A - \lambda I$ is invertible. This proves the required result. $\square$

Corollary 15.2.2 *Let operator matrix (1.1) be an upper triangular one, and $\tilde{V}(\lambda)$ be bounded for all regular λ of D. Then*

$$\sigma(A) = \cup_{k=1}^{n} \sigma(A_{kk}) = \sigma(D). \tag{2.3}$$

Similarly, let (1.1) be lower triangular and $\tilde{W}(\lambda)$ be bounded for all regular λ of D. Then (2.3) holds.

Indeed, let A be upper triangular. Then $\tilde{W}(\lambda) = 0$. Now the result is due to Theorem 15.2.1. The lower triangular case can be similarly considered.

This result shows that Theorem 15.2.1 is exact.

Lemma 15.2.3 *Let W and V be bounded operators and the condition*

$$\|(D - I_H \lambda)^{-1}\| \leq \Phi(\rho^{-1}(D, \lambda)) \ (\lambda \notin \sigma(D)) \tag{2.4}$$

hold, where $\Phi(y)$ is a continuous increasing function of $y \geq 0$ with the properties $\Phi(0) = 0$ and $\Phi(\infty) = \infty$. In addition, let z_0 be the unique positive root of the scalar equation

$$\sum_{j,k=1}^{n-1} \Phi^{k+j}(y)\|V\|^{j}\|W\|^{k} = 1. \tag{2.5}$$

Then the spectral variation of operator A defined by (1.1) with respect to D satisfies the inequality

$$sv_D(A) \leq \frac{1}{z_0}.$$

Proof: Due to (2.4)

$$\|\tilde{V}(\lambda)\| \leq \|V\|\Phi(\rho^{-1}(D, \lambda)), \|\tilde{W}(\lambda)\| \leq \|W\|\Phi(\rho^{-1}(D, \lambda)).$$

For any $\lambda \in \sigma(A)$ and $\lambda \notin \sigma(D)$, Theorem 15.2.1 implies

$$\sum_{j,k=1}^{n-1} \Phi^{k+j}(\rho^{-1}(D, \lambda))\|V\|^{k}\|W\|^{j} \geq 1.$$

Taking into account that Φ is increasing and comparing the latter inequality with (2.5), we have

$$\rho^{-1}(D, \lambda) \geq z_0$$

for any $\lambda \in \sigma(A)$. This proves the required result. $\square$

For instance, let $n = 2$. Then (2.5) takes the form

$$\Phi^2(y)\|V\|\|W\| = 1. \tag{2.6}$$

Hence it follows that

$$z_0 = \Psi(\frac{1}{\sqrt{\|V\|\|W\|}}),$$

where Ψ is the function inverse to Φ: $\Phi(\Psi(y)) = y$. Thus, in the case $n = 2$,

$$sv_D(A) \leq \frac{1}{\Psi(\frac{1}{\sqrt{\|V\|\|W\|}})}.$$

15.3   Operator Matrices with Normal Entries

Assume that H is an orthogonal sum of the same Hilbert spaces $E_k \equiv E$ ($k = 1, ..., n$) with norm $\|.\|_E$. Consider in H the operator matrix defined by (1.1), assuming that

$$A_{jj} = S_j \ (j = 1, ..., n), \tag{3.1}$$

where S_j are normal, unbounded in general operators in E_j, and

$$A_{jk} = \phi_k(S_j) \ (j \neq k; \ j, k = 1, ..., n), \tag{3.2}$$

where $\phi_k(s)$ are scalar-valued measurable functions of $s \in \mathbf{C}$. In addition, assume that

$$\alpha_{jk} \equiv \sup_{t \in \sigma(S_j)} |\phi_k(t)t^{-1}| < \infty. \tag{3.3}$$

Then $A_{jj}^{-1}A_{jk}$ are bounded normal operators with the norms

$$\|A_{jj}^{-1}A_{jk}\|_E = \alpha_{jk}.$$

Moreover, due to Lemma 15.1.4, with the notation

$$v_A := \max_j \sum_{k=j+1}^{n} \alpha_{jk} \text{ and } w_A := \max_j \sum_{k=1}^{j-1} \alpha_{jk}$$

we have

$$\|D^{-1}V\| \leq v_A \text{ and } \|D^{-1}W\| \leq w_A.$$

Now Corollary 15.1.3 implies

Lemma 15.3.1 *Let the conditions (3.1)-(3.3) and*

$$\sum_{j,k=1}^{n-1} v_A^k w_A^j < 1 \tag{3.4}$$

hold. Then operator (1.1) is invertible. In particular, let $v_A, w_A \neq 1$. Then (3.4) can be written in the form

$$v_A w_A \frac{(1 - v_A^{n-1})(1 - w_A^{n-1})}{(1 - v_A)(1 - w_A)} < 1. \tag{3.5}$$

Furthermore, under (3.1), (3.2) assume that for all regular λ of D, the relations

$$\tilde{v}(\lambda) := \max_j \sum_{k=j+1}^{n} \sup_{t \in \sigma(S_j)} |\phi_k(t)(\lambda - t)^{-1}| < \infty \text{ and}$$

$$\tilde{w}(\lambda) := \max_j \sum_{k=1}^{j-1} \sup_{t \in \sigma(S_j)} |\phi_k(t)(\lambda - t)^{-1}| < \infty. \tag{3.6}$$

are fulfilled. Due to Lemma 15.1.4 with the notation from (2.1), we have

$$\|\tilde{W}(\lambda)\| \leq \tilde{w}(\lambda) \text{ and } \|\tilde{V}(\lambda)\| \leq \tilde{v}(\lambda).$$

Now Theorem 15.2.1 gives.

Lemma 15.3.2 *Under conditions (3.1), (3.2) and (3.6), the spectrum of operator A defined by (1.1) lies in the union of the sets $\sigma(D)$ and*

$$\{\lambda \in \mathbf{C} : \sum_{j,k=1}^{n-1} \tilde{v}^k(\lambda)\tilde{w}^j(\lambda) \geq 1\}.$$

15.4   Operator Matrices with Bounded off Diagonal Entries

Again assume that H is an orthogonal sum of the same Hilbert spaces $E_k \equiv E$ ($k = 1, ..., n$). In addition, A_{jk} ($j \neq k$) are arbitrary bounded operators:

$$v_0 \equiv \|V\| < \infty, w_0 \equiv \|W\| < \infty. \tag{4.1}$$

Lemma 15.4.1 *Let the conditions (3.1),(4.1) and*

$$\sum_{j,k=1}^{n-1} \rho^{-k-j}(D,0)v_0^k w_0^j < 1$$

hold. Then operator (1.1) is invertible.

Proof: Since D is normal,

$$\|(D - \lambda I_H)^{-1}\| = \rho^{-1}(D,\lambda). \tag{4.2}$$

Therefore $\|D^{-1}\| = \rho^{-1}(D,0)$. Due to (4.1)

$$\|D^{-1}V\| \leq \|D^{-1}\|\|V\| \leq \rho^{-1}(D,0)v_0.$$

Similarly, $\|D^{-1}W\| \leq \rho^{-1}(D,0)w_0$. Now Corollary 15.1.3 implies the required result. $\square$

Lemma 15.4.2 *Under conditions (3.1) and (4.1), let z_1 be the unique non-negative root of the algebraic equation*

$$\sum_{j,k=0}^{n-2} z^{k+j} v_0^{n-j-1} w_0^{n-k-1} = z^{2(n-1)}. \tag{4.3}$$

Then the spectral variation of A with respect to D satisfies the inequality

$$sv_D(A) \le z_1.$$

Proof: With the substitution $y = 1/z$ equation (4.3) takes the form

$$\sum_{j,k=1}^{n-1} y^{k+j} v_0^j w_0^k = 1. \tag{4.4}$$

So under consideration equation (2.5) can be rewritten as (4.4). Due to Lemma 15.2.3 we arrive at the result. □

Due to Lemma 1.6.1, if

$$p(A) := \sum_{j,k=1}^{n-1} v_0^j w_0^k < 1, \tag{4.5}$$

then $z_1^{2(n-1)} \le p(A)$. So under (4.5), the previous lemma gives the inequality

$$sv_D(A) \le \sqrt[2(n-1)]{p(A)}. \tag{4.6}$$

Let us improve Lemma 15.4.1 in the case, when A_{jk} $(j \ne k)$ are Hilbert-Schmidt operators. Recall that $N_2(.)$ denotes the Hilbert-Schmidt norm. Clearly,

$$N_2^2(V) = \sum_{1 \le j < k \le n} N_2^2(A_{jk}),$$

and

$$N_2^2(W) = \sum_{1 \le k < j \le n} N_2^2(A_{jk}).$$

Lemma 15.4.3 *Under condition (3.1), let V and W be Hilbert-Schmidt operators and the inequality*

$$\sum_{j,k=1}^{n-1} (k!j!)^{-1/2} N_2^k(V) N_2^j(W) \rho^{-k-j}(D,0) < 1 \tag{4.7}$$

hold. Then operator A defined by (1.1) is invertible.

Proof: Due to (4.7) D^{-1} is bounded, $D^{-1}V$ and $D^{-1}W$ are Hilbert-Schmidt operators. In addition, according to (1.5), they are nilpotent. Due to Corollary 6.9.2,

$$\|(D^{-1}V)^k\| \le N_2^k(D^{-1}V)(k!)^{-1/2}, \|(D^{-1}W)^k\| \le N_2^k(D^{-1}W)(k!)^{-1/2}$$
$$(4.8)$$

for any natural $k < n$. But

$$N_2(D^{-1}V) \le \|D^{-1}\|N_2(V) = \rho^{-1}(D,0)N_2(V).$$

Similarly,

$$N_2(D^{-1}W) \le \rho^{-1}(D,0)N_2(W).$$

Now Theorem 15.1.1 implies the required result. $\square$

15.5   Operator Matrices with Hilbert-Schmidt Diagonal Operators

Again, let H be an orthogonal sum of Hilbert spaces E_k, $k = 1,...,n$. In addition, E_k are separable, condition (4.1) holds and the diagonal operators have the form

$$A_{jj} = I_E + K_j \text{ where } K_j \ (j = 1,...,n) \text{ are Hilbert-Schmidt operators .}$$
$$(5.1)$$

Due to Theorem 6.4.1, for any Hilbert-Schmidt operator K in H, we can write

$$\|(K-\lambda I)^{-1}\| \le G(K,\rho(\lambda,K)) \equiv \sum_{k=0}^{\infty} \frac{g^k(K)}{\sqrt{k!}\rho^{k+1}(K,\lambda)} \text{ for all regular } \lambda, \ (5.2)$$

where

$$G(K,y) \equiv \sum_{k=0}^{\infty} \frac{g^k(K)}{\sqrt{k!}y^{k+1}} \ (y > 0)$$

and

$$g(K) = (N_2^2(K) - \sum_{k=1}^{\infty} |\lambda_k(K)|^2)^{1/2} \ (K \in C_2).$$

If K is a normal operator: $KK^* = K^*K$, then $g(K) = 0$. The following relations are true:

$$g^2(K) \le N_2^2(K) - |Trace(K^2)|$$

and

$$g^2(K) \le \frac{1}{2}N_2^2(K^* - K) \ (K \in C_2)$$

(see Section 6.3). So due to (5.1) and (5.2)

$$\|(K_j - \lambda I)^{-1}\|_{E_j} \leq G(K_j, \rho(K_j, \lambda)) \tag{5.3}$$

and therefore with $\lambda = -1$ we get

$$\|D^{-1}\| = \max_j \|(I_E + K_j)^{-1}\|_{E_j} \leq b_0(D), \tag{5.4}$$

where

$$b_0(D) = \max_j G(K_j, \rho(K_j, -1)) = \max_j \sum_{k=0}^{\infty} \frac{g^k(K_j)}{\sqrt{k!}\rho^{k+1}(K_j, -1)}.$$

Lemma 15.5.1 *Let the conditions (4.1), (5.1) and*

$$\sum_{j,k=1}^{n-1} b_0^{k+j}(D)v_0^k w_0^j < 1 \tag{5.5}$$

be fulfilled. Then operator A defined by (1.1) is invertible.

Proof: Due to (4.1) and (5.4)

$$\|D^{-1}V\| \leq \|D^{-1}\|v_0 \leq b_0(D)v_0.$$

Similarly,

$$\|D^{-1}W\| \leq b_0(D)w_0.$$

Now Theorem 15.1.1 implies the required result. $\square$

Thanks to Theorem 15.1.1, we also get the following result.

Lemma 15.5.2 *Under conditions (4.1) and (5.1) the spectrum of operator A defined by (1.1), lies in the set*

$$\cup_{j=1}^n \Omega_j(\lambda),$$

where

$$\Omega_j(\lambda) = \{\lambda \in \mathbf{C} : \sum_{s,k=1}^{n-1} G^{s+k}(K_j, \rho(K_j, \lambda - 1))v_0^k w_0^s \geq 1\}.$$

In other words, for any $\mu \in \sigma(A)$, there are an integer j and a

$$\lambda(K_j) \in \sigma(K_j),$$

such that

$$\sum_{s,k=1}^{n-1} G^{s+k}(K_j, |\lambda(K_j) + 1 - \mu|)v_0^k w_0^s \geq 1.$$

Put

$$h(y, D) = \sum_{k=0}^{\infty} \frac{g_0^k y^{k+1}}{\sqrt{k!}},$$

where $g_0 = max_j g(K_j)$. Let us consider the scalar equation

$$\sum_{s,l=1}^{n-1} h^{s+l}(y, D) v_0^l w_0^s = 1. \tag{5.6}$$

Thanks to (5.3), one can write

$$\|(D - \lambda I_H)^{-1}\| = \max_j \|(I_E + K_j - \lambda I_E)^{-1}\| \leq h(\rho^{-1}(D, \lambda), D).$$

Now Lemma 15.2.3 yields

Theorem 15.5.3 *Let x_0 be the extreme right (unique positive) root of equation (5.6). Then the spectral variation of operator A defined by (1.1) with respect to D satisfies the inequality*

$$sv_D(A) \leq x_0^{-1}.$$

15.6   Example

Let $H = L^2([0, \pi], \mathbf{C}^n)$ be the Hilbert space of functions defined on $[0, \pi]$ with values in a Euclidean space $\mathbf{C}^n$ and the scalar product

$$(h, w)_H = \int_0^{\pi} (h(x), w(x))_{\mathbf{C}^n} dx,$$

where $(.,.)_{\mathbf{C}^n}$ is the scalar product in $\mathbf{C}^n$. Consider the operator A defined by the expression

$$Au(x) = -\frac{d}{dx} d_0(x) \frac{du(x)}{dx} + B_0(x)u(x) \quad (u \in Dom\,(A),\ 0 < x < \pi) \tag{6.1}$$

on the domain

$$Dom\,(A) = \{u \in H,\ u'' \in H,\ u(0) = u(\pi) = 0\,\}, \tag{6.2}$$

with continuous real $n \times n$-matrices

$$d_0(x) = diag\,[a_1(x), ..., a_n(x)],\, B_0(x) = (b_{jk}(x))_{j,k=1}^n,$$

where functions $a_j(x)$ are differentiable and positive:

$$\tilde{d}_j \equiv \min_x a_j(x) > 0. \tag{6.3}$$

Take $E = L^2([0, \pi], \mathbf{C}^1)$ and define operators A_{jk} by

$$A_{jj}v(x) = -\frac{d}{dx}a_j(x)\frac{dv(x)}{dx} + b_{jj}(x)v(x)$$

$$(v \in Dom\ (A_{jj}),\ \ 0 < x < \pi),$$

where

$$Dom\ (A_{jj}) = \{v \in E:\ v'' \in E,\ v(0) = v(\pi) = 0\ \}$$

and

$$A_{jk}v(x) = b_{jk}(x)v(x)\ \ (v \in E;\ \ 0 < x < \pi,\ j \neq k).$$

Assume that

$$\beta_j \equiv \inf_x b_{jj}(x) + \tilde{d}_j > 0\ (j = 1, ..., n). \tag{6.4}$$

Omitting simple calculations, we have

$$(A_{jj}v(x), v)_E = (a_j v', v')_E + (b_{jj}v, v)_E \geq \tilde{d}_j(v', v')_E + (b_{jj}v, v)_E \geq$$

$$\tilde{d}_j(v, v)_E + (b_{jj}v, v)_E = \beta_j(v, v)_E.$$

Consequently,

$$\|A_{jj}^{-1}\|_E \leq \beta_j^{-1}\ (j = 1, ..., n).$$

Clearly,

$$\|A_{jk}\|_E \leq \max_x |b_{jk}(x)|.$$

So

$$\|A_{jj}^{-1}A_{jk}\|_E \leq \beta_j^{-1}\max_x |b_{jk}(x)|.$$

With the notation

$$\tilde{v}_A \equiv (\sum_{j=1}^{n-1}\sum_{k=j+1}^{n}\beta_j^{-2}\max_x |b_{jk}(x)|^2)^{1/2}.$$

and

$$\tilde{w}_A \equiv (\sum_{j=2}^{n}\sum_{k=1}^{j-1}\beta_j^{-2}\max_x |b_{jk}(x)|^2)^{1/2}$$

we have

$$\|D^{-1}V\| \leq \tilde{v}_A,$$

and

$$\|D^{-1}W\| \leq \tilde{w}_A.$$

Now Corollary 15.1.3 yields

Proposition 15.6.1 *Let the conditions (6.3), (6.4) and*

$$\sum_{j,k=1}^{n-1} \tilde{v}_A^k \tilde{w}_A^j < 1 \tag{6.5}$$

hold. Then operator A defined by (6.1), (6.2) is invertible. In particular, let

$$\tilde{v}_A, \tilde{w}_A \neq 1.$$

Then (6.5) can be written in the form

$$\tilde{v}_A \tilde{w}_A \frac{(1 - \tilde{v}_A^{n-1})(1 - \tilde{w}_A^{n-1})}{(1 - \tilde{v}_A)(1 - \tilde{w}_A)} < 1.$$

In the case $n = 2$ one can write

$$\tilde{v}_A = \beta_1^{-1} \max_x |b_{12}(x)|, \ \tilde{w}_A = \beta_2^{-1} \max_x |b_{21}(x)|.$$

Inequality (6.5) takes the form

$$\max_x |b_{12}(x)| \ \max_x |b_{21}(x)| < \beta_1 \beta_2.$$

To investigate the spectrum of operator (6.1) assume for simplicity that

$$a_j \equiv const > 0, \ b_{jj} \equiv const \ (j = 1, 2, ...). \tag{6.6}$$

Then it is simple to check that the eigenvalues of A_{jj} are

$$\lambda_k(A_{jj}) = a_j k^2 + b_{jj} \ (k = 1, 2, ...).$$

Denote

$$v_b \equiv \left(\sum_{j=1}^{n-1} \sum_{k=j+1}^{n} \max_x |b_{jk}(x)|^2\right)^{1/2}$$

and

$$w_b \equiv \left(\sum_{j=1}^{n} \sum_{k=2}^{j-1} \max_x |b_{jk}(x)|^2\right)^{1/2}.$$

Clearly,

$$\|V\| \leq v_b, \ \|W\| \leq w_b.$$

Now Lemma 15.4.2 yields

Proposition 15.6.2 *Let z_2 be the unique non-negative root of the algebraic equation*

$$\sum_{j,k=0}^{n-2} z^{k+j} v_b^{n-j-1} w_b^{n-k-1} = z^{2(n-1)}. \tag{6.7}$$

Then the spectral variation of A with respect to D satisfies the inequality

$$sv_D(A) \leq z_2.$$

In other words for any $\mu \in \sigma(A)$, there are natural $k = 1, 2, \ldots$ and $j \leq n$, such that

$$|\mu - a_j k^2 - b_{jj}| \leq z_2.$$

In particular, if $n = 2$, then

$$z_2 = \sqrt{v_b w_b} = [\max_x |b_{12}(x)| \ \max_x |b_{21}(x)|]^{1/2}.$$

Certainly, instead of the ordinary differential operator, in (6.1) we can consider an elliptic one.

15.7   Notes

The spectrum of operator matrices and related problems were investigated in many works cf. (Kovarik, 1975, 1977 and 1980), (Kovarik and Sherif, 1985), (Gaur and Kovarik, 1991), (Stampli, 1964), (Davis, 1958) and references given therein. In particular, in the paper (Kovarik, 1975), the Gershgorin type bounds for spectra of operator matrices with bounded operator entries are derived. They generalize the well-known results for block-matrices (Varga, 1965), (Levinger and Varga, 1966). But the Gershgorin-type bounds give good results in the cases when the diagonal operators are dominant.

Theorem 15.1.1 improves the Gershgorin type bounds for operator matrices, which are close to triangular ones. Moreover, we also consider unbounded operators.

Proposition 15.6.2 on the bounds for the spectrum of a matrix differential operator supplements the well known results on differential operators, cf. (Edmunds and Evans, 1990), (Egorov and Kondratiev, 1996) and references therein.

The material in this chapter is taken from the paper (Gil', 2001).

References

[1] Davis, C. C. (1958). Separation of two subspaces. *Acta Sci. Math. (Szeged)* **19**, 172-187.

[2] Edmunds, D.E. and Evans W.D. (1990). *Spectral Theory and Differential Operators.* Clarendon Press, Oxford.

[3] Egorov, Y and Kondratiev, V. (1996). *Spectral Theory of Elliptic Operators.* Birkhäuser Verlag, Basel.

[4] Gaur A.K. and Kovarik, Z. V. (1991). Norms, states and numerical ranges on direct sums, *Analysis* **11**, 155-164.

[5] Gil', M. I. (2001). Invertibility conditions and bounds for spectra of operator matrices. *Acta Sci. Math*, **67/1**, 353-368

[6] Kato, T. (1966). *Perturbation Theory for Linear Operators*, Springer-Verlag. New York.

[7] Kovarik, Z. V. (1975). Spectrum localization in Banach spaces II, *Linear Algebra and Appl.* **12**, 223-229 .

[8] Kovarik, Z. V. (1977). Similarity and interpolation between projectors, *Acta Sci. Math. (Szeged)*, **39**, 341-351

[9] Kovarik, Z. V. (1980). Manifolds of frames of projectors, *Linear Algebra and Appl.* **31**, 151-158.

[10] Kovarik, Z. V. and Sherif, N. (1985). Perturbation of invariant subspaces, *Linear Algebra and Appl.* **64**, 93-113.

[11] Levinger, B.W. and Varga, R.S. (1966). Minimal Gershgorin sets II, *Pacific. J. Math.*, **17**, 199-210.

[12] Stampli, J. (1964), Sums of projectors, *Duke Math. J.*, **31**, 455-461.

[13] Varga, R.S. (1965). Minimal Gershgorin sets, *Pacific. J. Math.*, **15**, 719-729.

16. Hille - Tamarkin Integral Operators

In the present chapter, the Hille-Tamarkin integral operators on space $L^p[0,1]$ are considered. Invertibility conditions, estimates for the norm of the inverse operators and positive invertibility conditions are established. In addition, bounds for the spectral radius are suggested. Applications to nonselfadjoint differential operators and integro-differential ones are also discussed.

16.1 Invertibility Conditions

Recall that $L^p \equiv L^p[0,1]$ $(1 < p < \infty)$ is the space of scalar-valued functions defined on $[0,1]$ and equipped with the norm

$$|h|_{L^p} = [\int_0^1 |h(s)|^p ds]^{1/p}.$$

Everywhere below $\tilde{K}$ is a linear operator in L^p defined by

$$(\tilde{K}h)(x) = \int_0^1 K(x,s)h(s)ds \ (h \in L^p, x \in [0,1]), \tag{1.1}$$

where $K(x,s)$ is a scalar kernel defined on $[0,1]^2$ and having the property

$$M_p(K) \equiv [\int_0^1 [\int_0^1 |K(x,s)|^q ds]^{p/q} dx]^{1/p} < \infty \ (p^{-1} + q^{-1} = 1). \tag{1.2}$$

That is, $\tilde{K}$ is a Hille-Tamarkin operator (Pietsch, 1987, p. 245). Define the Volterra operators

$$(V_-h)(x) = \int_0^x K(x,s)h(s)ds \tag{1.3}$$

and

$$(V_+h)(x) = \int_x^1 K(x,s)h(s)ds. \qquad (1.4)$$

Set

$$M_p(V_-) \equiv [\int_0^1 (\int_0^t |K(t,s)|^q ds)^{p/q} dt]^{1/p},$$

$$M_p(V_+) \equiv [\int_0^1 (\int_t^1 |K(t,s)|^q ds)^{p/q} dt]^{1/p}$$

and

$$J_p^\pm \equiv \sum_{k=0}^\infty \frac{M_p^k(V_\pm)}{\sqrt[p]{k!}}.$$

Now we are in a position to formulate the main result of the chapter.

Theorem 16.1.1 *Let the conditions (1.2) and*

$$J_p^+ J_p^- < J_p^+ + J_p^- \qquad (1.5)$$

hold. Then operator $I - \tilde{K}$ *is boundedly invertible in* L^p *and the inverse operator satisfies the inequality*

$$|(I - \tilde{K})^{-1}|_{L^p} \le \frac{J_p^- J_p^+}{J_p^+ + J_p^- - J_p^+ J_p^-}. \qquad (1.6)$$

The proof of this theorem is presented in the next two sections.

Note that condition (1.5) is equivalent to the following one:

$$\theta(K) \equiv (J_p^+ - 1)(J_p^- - 1) < 1. \qquad (1.7)$$

Besides (1.6) takes the form

$$|(I - \tilde{K})^{-1}|_{L^p} \le \frac{J_p^- J_p^+}{1 - \theta(K)}. \qquad (1.8)$$

Due to Hölder's inequality, for arbitrary $a > 1$

$$\sum_{k=0}^\infty \frac{M_p^k(V_\pm)}{\sqrt[p]{k!}} \le$$

$$[\sum_{k=0}^\infty a^{-qk}]^{1/q} [\sum_{k=0}^\infty \frac{a^{pk} M_p^{kp}(V_\pm)}{k!}]^{1/p} = (1 - a^{-q})^{-1/q} e^{a^p M_p^p(V_\pm)/p}.$$

Take $a = 2^{1/p}$. Then

$$J_p^\pm \le m_p e^{2M_p^p(V_\pm)/p} \qquad (1.9)$$

where

$$m_p = (1 - 2^{-q/p})^{-1/q}.$$

Since,

$$M_p^p(K) = M_p^p(V_-) + M_p^p(V_+),\qquad (1.10)$$

we have

$$J_p^- J_p^+ \le m_p^2 e^{2M_p^p(K)/p}.$$

Now relation (1.9) and Theorem 16.1.1 imply

Corollary 16.1.2 *Let the conditions (1.2) and*

$$m_p e^{2M_p^p(K)/p} < e^{2M_p^p(V_-)/p} + e^{2M_p^p(V_+)/p}$$

hold. Then operator $I - \tilde{K}$ *is boundedly invertible in* L^p *and the inverse operator satisfies the inequality*

$$|(I - \tilde{K})^{-1}|_{L^p} \le \frac{m_p e^{2M_p^p(K)/p}}{e^{2M_p^p(V_-)/p} + e^{2M_p^p(V_+)/p} - m_p e^{2M_p^p(K)/p}}.$$

16.2 Preliminaries

Let X be a Banach space with a norm $\|.\|$. Recall that a linear operator $\tilde{V}$ in X is called a quasinilpotent one if

$$\lim_{n \to \infty} \sqrt[n]{\|\tilde{V}^n\|} = 0.$$

For a quasinilpotent operator $\tilde{V}$ in X, put

$$j(\tilde{V}) \equiv \sum_{k=0}^{\infty} \|\tilde{V}^k\|.$$

Lemma 16.2.1 *Let A be a bounded linear operator in X of the form*

$$A = I + V + W,\qquad (2.1)$$

where operators V and W are quasinilpotent. If, in addition, the condition

$$\theta_A \equiv \| \sum_{j,k=1}^{\infty} (-1)^{k+j} V^k W^j \| < 1 \qquad (2.2)$$

is fulfilled, then operator A is boundedly invertible and the inverse operator satisfies the inequality

$$\|A^{-1}\| \le \frac{j(V)j(W)}{1 - \theta_A}.$$

Proof: We have

$$A = I + V + W = (I + V)(I + W) - VW. \tag{2.3}$$

Since W and V are quasinilpotent, the operators, $I + V$ and $I + W$ are invertible:

$$(I + V)^{-1} = \sum_{k=0}^{\infty} (-1)^k V^k, \ (I + W)^{-1} = \sum_{k=0}^{\infty} (-1)^k W^k. \tag{2.4}$$

Thus,

$$A = I + V + W = (I + V)[I - (I + V)^{-1}VW(I + W)^{-1}](I + W) =$$

$$(I + V)(I - B_A)(I + W) \tag{2.5}$$

where

$$B_A = (I + V)^{-1}VW(I + W)^{-1}. \tag{2.6}$$

But according to (2.4)

$$V(I + V)^{-1} = \sum_{k=1}^{\infty} (-1)^{k-1} V^k, \ (I + W)^{-1} = \sum_{k=1}^{\infty} (-1)^{k-1} W^k. \tag{2.7}$$

So

$$B_A = \sum_{j,k=1}^{\infty} (-1)^{k+j} V^k W^j. \tag{2.8}$$

If (2.2) holds, then $\|B_A\| < 1$ and

$$\|(I - B_A)^{-1}\| \leq (1 - \theta_A)^{-1}.$$

So due to (2.5) $I + V + W$ is invertible. Moreover,

$$A^{-1} = (I + W)^{-1}(I - B_A)^{-1}(I + V)^{-1}. \tag{2.9}$$

But (2.4) implies

$$\|(I + W)^{-1}\| \leq j(W), \ \|(I + V)^{-1}\| \leq j(V).$$

Now the required inequality for A^{-1} follows from (2.9). □
Furthermore, take into account that by (2.7)

$$\|V(I + V)^{-1}\| \leq \sum_{k=1}^{\infty} \|V^k\| \leq j(V) - 1. \tag{2.10}$$

Similarly,

$$\|W(I + W)^{-1}\| \leq j(W) - 1. \tag{2.11}$$

Thus

$$\theta_A \le (j(W) - 1)(j(V) - 1).$$

So condition (2.2) is provided by the inequality

$$(j(W) - 1)(j(V) - 1) < 1.$$

The latter inequality is equivalent to the following one:

$$j(W)j(V) < j(W) + j(V) \tag{2.12}$$

Lemma 16.2.1 yields

Corollary 16.2.2 *Let* V, W *be quasinilpotent and condition (2.12) be fulfilled. Then operator* A *defined by (2.1) is boundedly invertible and the inverse operator satisfies the inequality*

$$\|A^{-1}\| \le \frac{j(V)j(W)}{j(W) + j(V) - j(W)j(V)}.$$

Let us turn now to integral operator $\tilde{K}$. Under condition (1.2), operators $V_\pm$ are quasinilpotent due to the well-known Theorem V.6.2 (Zabreiko, et al., 1968, p. 153). Now Corollary 16.2.2 yields.

Corollary 16.2.3 *Let the conditions (1.2) and*

$$j(V_+)j(V_-) < j(V_+) + j(V_-)$$

be fulfilled. Then $I - \tilde{K}$ *is boundedly invertible and the inverse operator satisfies the inequality*

$$|(I - \tilde{K})^{-1}|_{L^p} \le \frac{j(V_-)j(V_+)}{j(V_-) + j(V_+) - j(V_-)j(V_+)}.$$

16.3   Powers of Volterra Operators

Lemma 16.3.1 *Under condition (1.2), operator* V_- *defined by (1.3) satisfies the inequality*

$$|V_-^k|_{L^p} \le \frac{M_p^k(V_-)}{\sqrt[p]{k!}} \quad (k = 1, 2, ...). \tag{3.1}$$

Proof: Employing Hölder's inequality, we have

$$|V_- h|_{L^p}^p = \int_0^1 |\int_0^t K(t, s) h(s) ds|^p dt \le$$

$$\int_0^1 [\int_0^t |K(t, s)|^q ds]^{p/q} \int_0^t |h(s_1)|^p ds_1 dt.$$

Setting

$$w(t) = [\int_0^t |K(t,s)|^q ds]^{p/q}, \tag{3.2}$$

one can rewrite the latter relation in the form

$$|V_- h|_{L^p}^p \leq \int_0^1 w(s_1) \int_0^{s_1} |h(s_2)|^p ds_2 \, ds_1.$$

Using this inequality, we obtain

$$|V_-^k h|_{L^p}^p \leq \int_0^1 w(s_1) \int_0^{s_1} |V_-^{k-1} h(s_2)|^p ds_2 \, ds_1.$$

Once more apply Hölder's inequality :

$$|V_-^k h|_{L^p}^p \leq \int_0^1 w(s_1) \int_0^{s_1} w(s_2) \int_0^{s_2} |V_-^{k-2} h(s_3)|^p ds_3 \, ds_2 \, ds_1.$$

Repeating these arguments, we arrive at the relation

$$|V_-^k h|_{L^p}^p \leq \int_0^1 w(s_1) \int_0^{s_1} w(s_2) \ldots \int_0^{s_k} |h(s_{k+1})|^p ds_{k+1} \ldots ds_2 ds_1.$$

Taking

$$|h|_{L^p}^p = \int_0^1 |h(s)|^p ds = 1,$$

we get

$$|V_-^k|_{L^p}^p \leq \int_0^1 w(s_1) \int_0^{s_1} w(s_2) \ldots \int_0^{s_{k-1}} ds_k \ldots ds_2 ds_1. \tag{3.3}$$

It is simple to see that

$$\int_0^1 w(s_1) \ldots \int_0^{s_{k-1}} w(s_k) ds_k \ldots ds_1 =$$

$$\int_0^{\tilde{\mu}} \int_0^{z_1} \ldots \int_0^{z_{k-1}} dz_k dz_{k-1} \ldots dz_1 = \frac{\tilde{\mu}^k}{k!},$$

where

$$z_k = z_k(s_k) \equiv \int_0^{s_k} w(s) ds$$

and

$$\tilde{\mu} = \int_0^1 w(s) ds.$$

Thus (3.3) gives

$$|V_-^k|_{L^p}^p \leq \frac{(\int_0^1 w(s) ds)^k}{k!}.$$

But according to (3.2)

$$\tilde{\mu} = \int_0^1 w(s)ds = M_p^p(V_-).$$

Therefore,

$$|V_-^k|_{L^p}^p \le \frac{M^{pk}(V_-)}{k!}.$$

As claimed. $\square$

Similarly, the inequality

$$|V_+^k|_{L^p} \le \frac{M_p^k(V_+)}{\sqrt[p]{k!}} \tag{3.4}$$

can be proved.

The assertion of Theorem 16.1.1 follows from Corollary 16.2.3 and relations (3.1), (3.4).

16.4   Spectral Radius of a Hille - Tamarkin Operator

Set

$$J_p^{\pm}(z) \equiv \sum_{k=0}^{\infty} \frac{z^k M_p^k(V_{\pm})}{\sqrt[p]{k!}} \quad (z \ge 0).$$

So $J_p^{\pm} = J_p^{\pm}(1)$. Clearly,

$$\lambda I - \tilde{K} = \lambda(I - \lambda^{-1}\tilde{K}) \ (\lambda \ne 0).$$

Consequently, if

$$J_p^+(|\lambda|^{-1})J_p^-(|\lambda|^{-1}) < J_p^+(|\lambda|^{-1}) + J_p^-(|\lambda|^{-1}),$$

then due to Theorem 16.1.1, $\lambda I - \tilde{K}$ is boundedly invertible. We thus get

Lemma 16.4.1 *Under condition (1.2), any point $\lambda \ne 0$ of the spectrum $\sigma(\tilde{K})$ of operator $\tilde{K}$ satisfies the inequality*

$$J_p^+(|\lambda|^{-1})J_p^-(|\lambda|^{-1}) \ge J_p^+(|\lambda|^{-1}) + J_p^-(|\lambda|^{-1}). \tag{4.1}$$

Let $r_s(\tilde{K}) = \sup |\sigma(\tilde{K})|$ be the spectral radius of $\tilde{K}$. Then (4.1) yields

$$J_p^+(r_s^{-1}(\tilde{K}))J_p^-(r_s^{-1}(\tilde{K})) \ge J_p^+(r_s^{-1}(\tilde{K})) + J_p^-(r_s^{-1}(\tilde{K})). \tag{4.2}$$

Note that according to (1.9) and (4.2) we have

$$m_p exp\left[\frac{2(M_p^p(V_-) + M_p^p(V_+))}{r_s^p(\tilde{K})p}\right] \ge exp\left[\frac{2M_p^p(V_-)}{r_s^p(\tilde{K})p}\right] + exp\left[\frac{2M_p^p(V_+)}{r_s^p(\tilde{K})p}\right].$$

Theorem 16.4.2 *Under condition (1.2), let $V_- \neq 0$ and $V_+ \neq 0$. Then the equation*

$$J_p^+(z)J_p^-(z) = J_p^+(z) + J_p^-(z) \tag{4.3}$$

has a unique positive zero $z(K)$. Moreover, the inequality $r_s(\tilde{K}) \leq z^{-1}(K)$ is valid.

Proof: Equation (4.3) is equivalent to the following one:

$$(J_p^+(z) - 1)(J_p^-(z) - 1) = 1. \tag{4.4}$$

Clearly, this equation has a unique positive root. In addition, (4.2) is equivalent to the relation

$$(J_p^+(r_s^{-1}(\tilde{K})) - 1)(J_p^-(r_s^{-1}(\tilde{K})) - 1) \geq 1.$$

Hence the result follows, since the left part of equation (4.4) monotonically increases. $\square$

Rewrite (4.4) as

$$\sum_{k=1}^{\infty} \frac{z^k M_p^k(V_-)}{\sqrt[p]{k!}} \sum_{j=1}^{\infty} \frac{z^j M_p^j(V_+)}{\sqrt[p]{j!}} = 1$$

Or

$$\sum_{k=2}^{\infty} b_k z^k = 1$$

with

$$b_k = \sum_{j=1}^{k-1} \frac{M_p^{k-j}(V_-)M_p^j(V_+)}{\sqrt[p]{j!(k-j)!}} \quad (k = 2, 3, ...).$$

Due to Lemma 8.3.1, with the notation

$$\delta(K) = 2 \max_{j=2,3,...} \sqrt[j]{b_j},$$

we get $z(K) \geq \delta^{-1}(K)$. Now Theorem 16.4.2 yields

Corollary 16.4.3 *Under condition (1.2), the inequality $r_s(\tilde{K}) \leq \delta(K)$ is true.*

Theorem 16.4.2 and Corollary 16.4.3 are exact: if either $V_- \to 0$, or $V_+ \to 0$, then $z(K) \to \infty$, $\delta(K) \to 0$.

16.5   Nonnegative Invertibility

We will say that $h \in L^p$ is nonnegative if $h(t)$ is nonnegative for almost all $t \in [0,1]$; a linear operator A in L^p is nonnegative if Ah is nonnegative for each nonnegative $h \in L^p$.

Theorem 16.5.1 *Let the conditions (1.2), (1.5) and*

$$K(t,s) \geq 0 \ (0 \leq t, s \leq 1) \tag{5.1}$$

hold. Then operator $I - \tilde{K}$ is boundedly invertible and the inverse operator is nonnegative. Moreover,

$$(I - \tilde{K})^{-1} \geq I. \tag{5.2}$$

Proof: Relation (2.9) with $A = I - \tilde{K}$, $W = V_-$ and $V = V_+$ implies

$$(I - \tilde{K})^{-1} = (I - V_+)^{-1}(I - B_K)^{-1}(I - V_-)^{-1} \tag{5.3}$$

where

$$B_K = (I - V_+)^{-1}V_+V_-(I - V_-)^{-1}.$$

Moreover, by (5.1) we have $V_\pm \geq 0$. So due to (2.4), $(I - V_\pm)^{-1} \geq 0$ and $B_K \geq 0$. Relations (2.7) and (2.8) according to (2.10) and (2.11) imply

$$|B_K|_{L^p} \leq (J_p(V_-) - 1)(J_p(V_+) - 1),$$

since $j(V_\pm) \leq J_p(V_\pm)$. But (1.5) is equivalent to (1.7). We thus get $|B_K|_{L^p} < 1$. Consequently,

$$(I - B_K)^{-1} = \sum_{k=0}^{\infty} B_K^k \geq 0.$$

Now (5.3) implies the inequality $(I - \tilde{K})^{-1} \geq 0$. Since $I - \tilde{K} \leq I$, we have inequality (5.2). □

16.6   Applications

16.6.1   A nonselfadjoint differential operator

Consider a differential operator A defined by

$$(Ah)(x) = -\frac{d^2h(x)}{dx^2} + g(x)\frac{dh(x)}{dx} + w(x)h(x) \ (0 < x < 1, \ h \in Dom\ (A)) \tag{6.1}$$

on the domain

$$Dom\ (A) = \{h \in L^p : \ h'' \in L^p + \text{ boundary conditions }\}. \tag{6.2}$$

In addition,

$$\text{the coefficients } g, w \in L^p \text{ and are complex, in general.} \tag{6.3}$$

Let an operator S be defined on $Dom\ (A)$ by

$$(Sh)(x) = -h''(x),\ h \in Dom\ (A).$$

It is ssumed that S has the Green function $G(t, s)$. So that,

$$(S^{-1}h)(x) \equiv \int_0^1 G(x, s)h(s)ds \in Dom\ (A)$$

for any $h \in L^p$. Besides, the derivative of the Green function in x satisfies the inequality

$$vrai\ sup_x \int_0^1 |G_x(x, s)|^q ds < \infty. \tag{6.4}$$

Thus, $A = (I - \tilde{K})S$, where

$$(\tilde{K}h)(x) = -(g(x)\frac{d}{dx} + w(x)) \int_0^1 G(x, s)h(s)ds = \int_0^1 K(x, s)h(s)ds$$

with

$$K(x, s) = -g(x)G_x(x, s) - w(x)G(x, s). \tag{6.5}$$

We have

$$\int_0^1 [\int_0^1 |g(x)G_x(x, s)|^q ds]^{p/q}dx =$$

$$\int_0^1 |g(x)|^p [\int_0^1 |G_x(x, s)|^q ds]^{p/q}dx < \infty.$$

Similarly,

$$\int_0^1 [\int_0^1 |w(x)G(x, s)|^q ds]^{p/q}dx =$$

$$\int_0^1 |w(x)|^p [\int_0^1 |G(x, s)|^q ds]^{p/q}dx < \infty.$$

Thus, condition (1.2) holds. Take into account that by Hölder's inequality

$$|S^{-1}h|_{L^p} = [\int_0^1 |\int_0^1 G(x, s)h(s)ds|^p dx]^{1/p} \leq b_p(S)|h|_{L^p}$$

where

$$b_p(S) = [\int_0^1 (\int_0^1 |G(x, s)|^q ds)^{p/q}dx]^{1/p}.$$

Since

$$A^{-1} = S^{-1}(I - \tilde{K})^{-1},$$

Theorem 16.1.1 immediately implies the following result:

Proposition 16.6.1 *Under (6.3)-(6.5), let condition (1.5) hold. Then operator A defined by (6.1), (6.2) is boundedly invertible in L^p. In addition,*

$$|A^{-1}|_{L^p} \leq \frac{b_p(S)J_p^- J_p^+}{J_p^+ + J_p^- - J_p^+ J_p^-}.$$

16.6.2 An integro-differential operator

On domain (6.2), let us consider the operator

$$(Eu)(x) = -\frac{d^2u(x)}{dx^2} + \int_0^1 K_0(x,s)u(s)ds \ (u \in Dom\ (A),\ 0 < x < 1),\ (6.6)$$

where K_0 is a kernel with the property

$$\int_0^1 \int_0^1 |K_0(x,s)|^p ds\ dx < \infty. \tag{6.7}$$

Let S be the same as in the previous subsection. Then we can write $E = (I - \tilde{K})S$ where $\tilde{K}$ is defined by (1.1) with

$$K(x,s) = -\int_0^1 K_0(x,x_1)G(x_1,s)dx_1 \tag{6.8}$$

So if $I - \tilde{K}$ is invertible, then E is invertible as well. By Hölder's inequality

$$\int_0^1 [\int_0^1 |\int_0^1 K_0(x,x_1)G(x_1,s)dx_1|^q ds]^{p/q}dx \le$$

$$\int_0^1 \int_0^1 |K_0(x,x_1)|^p dx_1 dx \ [\int_0^1 |G(x_1,s)|^q dx_1 ds]^{p/q}.$$

That is, condition (1.2) holds. Since

$$E^{-1} = S^{-1}(I - \tilde{K})^{-1},$$

Theorems 16.1.1 and 16.5.1 yield

Proposition 16.6.2 *Under (6.4), (6.7) and (6.8), let condition (1.5) hold. Then operator E defined by (6.6), (6.2) is boundedly invertible in L^p and*

$$|E^{-1}|_{L^p} \le \frac{b_p(S)J_p^- J_p^+}{J_p^+ + J_p^- - J_p^+ J_p^-}.$$

If, in addition, $G \ge 0$ and $K_0 \le 0$, then E^{-1} is positive. Moreover,

$$(E^{-1}h)(x) \ge (S^{-1}h)(x) = \int_0^1 G(x,s)h(s)ds$$

for any nonnegative $h \in L^p$.

16.7   Notes

A lot of papers and books are devoted to the spectrum of Hille-Tamarkin integral operators. Mainly, the distributions of the eigenvalues are considered, cf. (Diestel et al., 1995), (König, 1986), (Pietsch, 1987) and references therein.

Theorem 16.4.2 and Corollary 16.4.3 improve the well-known estimate

$$r_s(\tilde{K}) \leq \sup_x \int_0^1 |K(x,s)|ds$$

(Krasnosel'skii et al, 1989, Theorem 16.2) for operators, which are close to Volterra ones.

The results of Section 16.6 supplement the well-known results on the spectra of differential operators, cf. (Edmunds and Evans, 1990), (Egorov and Kondratiev, 1996), (Locker, 1999) and references therein.

The material in this chapter is taken from the paper (Gil', 2002).

References

[1] Diestel, D., Jarchow, H, Tonge, A. (1995), *Absolutely Summing Operators*, Cambridge University Press, Cambridge.

[2] Edmunds, D.E. and Evans W.D. (1990). *Spectral Theory and Differential Operators*. Clarendon Press, Oxford.

[3] Egorov, Y. and Kondratiev, V. (1996). *Spectral Theory of Elliptic Operators*. Birkhäuser Verlag, Basel.

[4] Gil', M.I. (2002). Invertibility and positive invertibility of Hille-Tamarkin integral operators, *Acta Math. Hungarica*, **95 (1-2)** 39-53.

[5] König, H. (1986). *Eigenvalue Distribution of Compact Operators*, Birkhäuser Verlag, Basel- Boston-Stuttgart.

[6] Locker, J. (1999). *Spectral Theory of Non-Self-Adjoint Two Point Differential Operators.*, Amer. Math. Soc., Mathematical Surveys and Monographs, Volume 73.

[7] Krasnosel'skii, M. A., J. Lifshits, and A. Sobolev (1989). *Positive Linear Systems. The Method of Positive Operators*, Heldermann Verlag, Berlin.

[8] Pietsch, A. (1987). *Eigenvalues and s-Numbers*, Cambridge University Press, Cambridge.

[9] Zabreiko, P.P., Koshelev A.I., Krasnosel'skii, M. A., Mikhlin, S.G., Rakovshik, L.S. and B.Ya. Stetzenko (1968). *Integral Equations*, Nauka, Moscow. In Russian

17. Integral Operators in Space L^∞

In the present chapter integral operators in space $L^\infty[0,1]$ are considered. Invertibility conditions, estimates for the norm of the inverse operators and positive invertibility conditions are established. In addition, bounds for the spectral radius are suggested. Applications to nonselfadjoint differential operators and integro-differential ones are also discussed.

17.1 Invertibility Conditions

Recall that $L^\infty \equiv L^\infty[0,1]$ is the space of scalar-valued functions defined on $[0,1]$ and equipped with the norm

$$|h|_{L^\infty} = ess \sup_{x \in [0,1]} |h(x)| \ (h \in L^\infty).$$

Everywhere in this chapter $\tilde{K}$ is a linear operator in L^∞ defined by

$$(\tilde{K}h)(x) = \int_0^1 K(x,s)h(s)ds \ (h \in L^\infty, x \in [0,1]), \tag{1.1}$$

where $K(x,s)$ is a scalar kernel defined on $[0,1]^2$ and having the property

$$\int_0^1 ess \sup_{x \in [0,1]} |K(x,s)|ds < \infty. \tag{1.2}$$

Define the Volterra operators

$$(V_-h)(x) = \int_0^x K(x,s)h(s)ds \tag{1.3}$$

and

$$(V_+ h)(x) = \int_x^1 K(x,s)h(s)ds. \tag{1.4}$$

Set

$$w_-(s) \equiv ess \sup_{0 \le s \le x \le 1} |K(x,s)|,$$

$$w_+(s) \equiv ess \sup_{0 \le x \le s \le 1} |K(x,s)|$$

and

$$M_\infty(V_\pm) \equiv \int_0^1 w_\pm(s)ds.$$

Now we are in a position to formulate the main result of the chapter.

Theorem 17.1.1 *Let the conditions (1.2) and*

$$e^{M_\infty(V_-)+M_\infty(V_+)} < e^{M_\infty(V_+)} + e^{M_\infty(V_-)} \tag{1.5}$$

hold. Then operator $I - \tilde{K}$ is boundedly invertible in L^∞ and the inverse operator satisfies the inequality

$$|(I - \tilde{K})^{-1}|_{L^\infty} \le \frac{e^{M_\infty(V_-)+M_\infty(V_+)}}{e^{M_\infty(V_+)} + e^{M_\infty(V_-)} - e^{M_\infty(V_-)+M_\infty(V_+)}}. \tag{1.6}$$

The proof of this theorem is presented in the next section.

Note that condition (1.5) is equivalent to the following one:

$$\theta(K) \equiv (e^{M_\infty(V_+)} - 1)(e^{M_\infty(V_-)} - 1) < 1. \tag{1.7}$$

Besides (1.6) takes the form

$$|(I - \tilde{K})^{-1}|_{L^\infty} \le \frac{e^{M_\infty(V_-)+M_\infty(V_+)}}{1 - \theta(K)}. \tag{1.8}$$

17.2   Proof of Theorem 17.1.1

Under condition (1.2), operators $V_\pm$ are quasinilpotent due to the well-known Theorem V.6.2 (Zabreiko, et al., 1968, p. 153). Now Corollary 16.2.2 yields.

Lemma 17.2.1 *With the notation*

$$j(V_\pm) \equiv \sum_{k=0}^{\infty} |V_\pm^k|_{L^\infty},$$

let the conditions (1.2) and

$$j(V_+)j(V_-) < j(V_+) + j(V_-)$$

be fulfilled. Then $I - \tilde{K}$ *is boundedly invertible in* L^∞ *and the inverse operator satisfies the inequality*

$$|(I - \tilde{K})^{-1}|_{L^\infty} \leq \frac{j(V_-)j(V_+)}{j(V_-) + j(V_+) - j(V_-)j(V_+)}.$$

Lemma 17.2.2 *Under condition (1.2), operator* V_- *defined by (1.3) satisfies the inequality*

$$|V_-^k|_{L^\infty} \leq \frac{M_\infty^k(V_-)}{k!} \quad (k = 1, 2, ...). \tag{2.1}$$

Proof: We have

$$|V_- h|_{L^\infty} = ess\ sup_{x \in [0,1]} | \int_0^x K(x, s)h(s)ds| \leq \int_0^1 w_-(s)|h(s)|ds.$$

Repeating these arguments, we arrive at the relation

$$|V_-^k h|_{L^\infty} \leq \int_0^1 w_-(s_1) \int_0^{s_1} w_-(s_2) \ldots \int_0^{s_k} |h(s_k)|ds_k \ldots ds_2 ds_1.$$

Taking $|h|_{L^\infty} = 1$, we get

$$|V_-^k|_{L^\infty} \leq \int_0^1 w_-(s_1) \int_0^{s_1} w_-(s_2) \ldots \int_0^{s_{k-1}} ds_k \ldots ds_2 ds_1. \tag{2.2}$$

It is simple to see that

$$\int_0^1 w_-(s_1) \ldots \int_0^{s_{k-1}} w_-(s_k)ds_k \ldots ds_1 =$$

$$\int_0^{\tilde{\mu}} \int_0^{z_1} \ldots \int_0^{z_{k-1}} dz_k dz_{k-1} \ldots dz_1 = \frac{\tilde{\mu}^k}{k!},$$

where

$$z_j = z_k(s_j) \equiv \int_0^{s_j} w_-(s)ds \quad (j = 1, ..., k)$$

and

$$\tilde{\mu} = \int_0^1 w_-(s)ds.$$

Thus (2.2) gives

$$|V_-^k|_{L^\infty} \leq \frac{(\int_0^1 w_-(s)ds)^k}{k!} = \frac{M_\infty^k(V_-)}{k!}.$$

As claimed. $\square$

Similarly, the inequality

$$|V_+^k|_{L^\infty} \leq \frac{M_\infty^k(V_+)}{k!} \ (k = 1, 2, ...) \tag{2.3}$$

can be proved.

Relations (2.1) and (2.3) imply

$$|(I - V_\pm)^{-1}|_{L^\infty} \leq j(V_\pm) \leq e^{M_\infty(V_\pm)}. \tag{2.4}$$

The assertion of Theorem 17.1.1 follows from Lemma 17.2.1 and relations (2.4).

17.3   The Spectral Radius

Clearly,

$$\lambda I - \tilde{K} = \lambda(I - \lambda^{-1}\tilde{K}) \ (\lambda \neq 0).$$

Consequently, if

$$e^{(M_\infty(V_-) + M_\infty(V_+))|\lambda|^{-1}} < e^{|\lambda|^{-1}M_\infty(V_+)} + e^{|\lambda|^{-1}M_\infty(V_-)},$$

then due to Theorem 17.1.1, $\lambda I - \tilde{K}$ is boundedly invertible. We thus get

Lemma 17.3.1 *Under condition (1.2), any point $\lambda \neq 0$ of the spectrum $\sigma(\tilde{K})$ of operator $\tilde{K}$ satisfies the inequality*

$$e^{(M_\infty(V_-) + M_\infty(V_+))|\lambda|^{-1}} \geq e^{|\lambda|^{-1}M_\infty(V_+)} + e^{|\lambda|^{-1}M_\infty(V_-)}. \tag{3.1}$$

Let $r_s(\tilde{K})$ be the spectral radius of $\tilde{K}$. Then (3.1) yields

$$e^{r_s^{-1}(\tilde{K})(M_\infty(V_-) + M_\infty(V_+))} \geq e^{r_s^{-1}(\tilde{K})M_\infty(V_+)} + e^{r_s^{-1}(\tilde{K})M_\infty(V_-)}. \tag{3.2}$$

Clearly, if $V_+ = 0$ or (and) $V_- = 0$, then $r_s(\tilde{K}) = 0$.

Theorem 17.3.2 *Under condition (1.2), let $V_+ \neq 0$, $V_- \neq 0$. Then the equation*

$$e^{(M_\infty(V_-) + M_\infty(V_+))z} = e^{zM_\infty(V_+)} + e^{zM_\infty(V_-)} \ (z \geq 0) \tag{3.3}$$

has a unique positive zero $z(K)$. Moreover, the inequality $r_s(\tilde{K}) \leq z^{-1}(K)$ is valid.

Proof: Equation (3.3) is equivalent to the following one:

$$(e^{M_\infty(V_+)z} - 1)(e^{zM_\infty(V_-)} - 1) = 1. \tag{3.4}$$

In addition, (3.2) is equivalent to the relation

$$(e^{r_s^{-1}(\tilde{K})M_\infty(V_+)} - 1)(e^{r_s^{-1}(\tilde{K})M_\infty(V_-)} - 1) \geq 1.$$

Hence, the result follows, since the left part of equation (3.4) monotonically increases. □

From (3.3) it follows that

$$e^{(M_\infty(V_-)+M_\infty(V_+))z} \geq 2$$

and

$$e^{z(M_\infty(V_+)-M_\infty(V_-))} = e^{M_\infty(V_+)z} - 1 \geq$$

$$exp\,[ln\,2\,M_\infty(V_+)(M_\infty(V_-)+M_\infty(V_+))^{-1}] - 1.$$

Thus with the notation

$$\delta_\infty(K) = \frac{M_\infty(V_+) - M_\infty(V_-)}{ln\,[\,exp\,(\frac{M_\infty(V_+)ln\,2}{M_\infty(V_-)+M_\infty(V_+)}) - 1]} \tag{3.5}$$

we have

$$z(K) \geq \delta_\infty^{-1}(K), \tag{3.6}$$

provided

$$M_\infty(V_+) < M_\infty(V_-). \tag{3.7}$$

Clearly, in (3.5) we can exchang the places of V_- and V_+. Now Theorem 17.3.2 yields

Corollary 17.3.3 *Under conditions (1.2) and (3.7), the inequality $r_s(\tilde{K}) \leq \delta_\infty(K)$ is true.*

17.4   Nonnegative Invertibility

We will say that $h \in L^\infty$ is nonnegative if $h(t)$ is nonnegative for almost all $t \in [0,1]$; a linear operator A in L^∞ is nonnegative if Ah is nonnegative for each nonnegative $h \in L^\infty$. Recall that I is the identity operator.

Theorem 17.4.1 *Let the conditions (1.2), (1.5) and*

$$K(t,s) \geq 0\ (0 \leq t, s \leq 1) \tag{4.1}$$

hold. Then operator $I - \tilde{K}$ is boundedly invertible and the inverse operator is nonnegative. Moreover,

$$(I - \tilde{K})^{-1} \geq I. \tag{4.2}$$

Proof: Relation (2.9) from Section 16.2 with $A = I - \tilde{K}$, $W = V_-$ and $V = V_+$ implies

$$(I - \tilde{K})^{-1} = (I - V_+)^{-1}(I - B_K)^{-1}(I - V_-)^{-1} \tag{4.3}$$

where

$$B_K = (I - V_+)^{-1}V_+V_-(I - V_-)^{-1}.$$

Moreover, by (4.1) we have $V_\pm \geq 0$. So $(I - V_\pm)^{-1} \geq 0$ and $B_K \geq 0$. Relations (2.4) give us the inequalities

$$|(I - V_\pm)^{-1}V_\pm|_{L^\infty} \leq e^{M_\infty(V_\pm)} - 1.$$

Consequently,

$$|B_K|_{L^\infty} \leq (e^{M_\infty(V_+)} - 1)(e^{M_\infty(V_-)} - 1).$$

But (1.5) is equivalent to (1.7). We thus get $|B_K|_{L^\infty} < 1$. Consequently,

$$(I - B_K)^{-1} = \sum_{k=0}^{\infty} B_K^k \geq 0.$$

Now (4.3) implies the inequality $(I - \tilde{K})^{-1} \geq 0$. In addition, since $I - \tilde{K} \leq I$, we have inequality (4.2). $\square$

17.5 Applications

17.5.1 A nonselfadjoint differential operator

Consider a differential operator A defined by

$$(Ah)(x) = -\frac{d^2h(x)}{dx^2} + g(x)\frac{dh(x)}{dx} + m(x)h(x) \ (0 < x < 1, \ h \in Dom\ (A)\ ) \tag{5.1}$$

on the domain

$$Dom\ (A) = \{h \in L^\infty,\ h'' \in L^\infty + \text{ some boundary conditions }\} \tag{5.2}$$

In addition,

the coefficients $g, w \in L^\infty$ and are complex, in general. $\tag{5.3}$

Let an operator S be defined on $Dom\ (A)$ by

$$(Sh)(x) = -h''(x),\ h \in Dom\ (A).$$

It is assumed that S has the Green function $G(t, s)$. So that,

$$(S^{-1}h)(x) \equiv \int_0^1 G(x, s)h(s)ds \in Dom\ (A)$$

for any $h \in L^\infty$, and the derivative of G in x satisfies the condition

$$\int_0^1 \sup_x |G_x(x, s)| ds < \infty. \tag{5.4}$$

Put

$$b_\infty(S) := \int_0^1 \sup_x |G(x, s)| ds.$$

We have

$$A = (I - \tilde{K})S,$$

where

$$(\tilde{K}h)(x) = -(g(x)\frac{d}{dx} + m(x))\int_0^1 G(x, s)h(s)ds = \int_0^1 K(x, s)h(s)ds$$

with

$$K(x, s) = -g(x)G_x(x, s) - m(x)G(x, s). \tag{5.5}$$

According to (5.3) and (5.4), condition (1.2) holds. Take into account that

$$|S^{-1}h|_{L^\infty} \le b_\infty(S)|h|_{L^\infty}.$$

Since

$$A^{-1} = S^{-1}(I - \tilde{K})^{-1},$$

Theorem 17.1.1 immediately implies the following result:

Proposition 17.5.1 *Under (5.3)-(5.5), let condition (1.5) hold. Then operator A defined by (5.1), (5.2) is boundedly invertible in L^∞. In addition,*

$$|A^{-1}|_{L^\infty} \le \frac{b_\infty(S)e^{M_\infty(V_-)+M_\infty(V_+)}}{e^{M_\infty(V_+)} + e^{M_\infty(V_-)} - e^{M_\infty(V_-)+M_\infty(V_+)}}.$$

17.5.2   An integro-differential operator

On domain (5.2), let us consider the operator

$$(Eu)(x) = -\frac{d^2u(x)}{dx^2} + \int_0^1 K_0(x, s)u(s)ds \ (u \in Dom\ (A),\ 0 < x < 1), \tag{5.6}$$

where K_0 is a kernel with the property

$$ess\ sup_x \int_0^1 |K_0(x, s)|\ ds < \infty. \tag{5.7}$$

Let S and G be the same as in the previous subsection. Then we can write $E = (I - \tilde{K})S$ where $\tilde{K}$ is defined by (1.1) with

$$K(x, s) = -\int_0^1 K_0(x, x_1)G(x_1, s)dx_1 \tag{5.8}$$

So if $I - \tilde{K}$ is invertible, then E is invertible as well. Clearly, under (5.4) and (5.7), condition (1.2) holds. Since

$$E^{-1} = S^{-1}(I - \tilde{K})^{-1},$$

Theorems 17.1.1 and 17.4.1 yield

Proposition 17.5.2 *Under (5.4), (5.7) and (5.8), let condition (1.5) hold. Then operator E defined by (5.6), (5.2) is boundedly invertible in L^∞ and*

$$|E^{-1}|_{L^\infty} \leq \frac{b_\infty(S)e^{M_\infty(V_-)+M_\infty(V_+)}}{e^{M_\infty(V_+)} + e^{M_\infty(V_-)} - e^{M_\infty(V_-)+M_\infty(V_+)}}.$$

If, in addition, $G \geq 0$ and $K_0 \leq 0$, then E^{-1} is positive. Moreover,

$$(E^{-1}h)(x) \geq (S^{-1}h)(x) = \int_0^1 G(x,s)h(s)ds$$

for any nonnegative $h \in L^\infty$.

17.6   Notes

The present chapter is based on the paper (Gil', 2001).

About well-known results on the spectrum of integral operators on L^∞, see, for instance, the books (Diestel et al., 1995), (König, 1986), (Krasnosel'skii et al., 1989), (Pietsch, 1987) and references therein.

References

[1] Diestel, D., Jarchow, H, Tonge, A. (1995), *Absolutely Summing Operators*, Cambridge University Press, Cambridge.

[2] Gil', M.I. (2001). Invertibility and positive invertibility conditions of integral operators in L^∞, *J. of Integral Equations and Appl.* **13** , 1-14.

[3] König, H. (1986). *Eigenvalue Distribution of Compact Operators*, Birkhäuser Verlag, Basel- Boston-Stuttgart.

[4] Krasnosel'skii, M. A., J. Lifshits, and A. Sobolev (1989). *Positive Linear Systems. The Method of Positive Operators*, Heldermann Verlag, Berlin.

[5] Pietsch, A. (1987). *Eigenvalues and s-Numbers*, Cambridge University Press, Cambridge.

[6] Zabreiko, P.P., A.I. Koshelev, M. A. Krasnosel'skii, S.G. Mikhlin, L.S. Rakovshik, B.Ya. Stetzenko (1968). *Integral Equations*, Nauka, Moscow. In Russian

18. Hille - Tamarkin Matrices

In the present chapter we investigate infinite matrices, whose off diagonal parts are the Hille-Tamarkin matrices. Invertibility conditions and estimates for the norm of the inverse matrices are established. In addition, bounds for the spectrum are suggested. In particular, estimates for the spectral radius are derived.

18.1 Invertibility Conditions

Everywhere in this chapter

$$A = (a_{jk})_{j,k=1}^{\infty}$$

is an infinite matrix with the entries a_{jk} $(j, k = 1, 2, ...)$. Besides, V_+, V_- and D denote the strictly upper triangular, strictly lower triangular, and diagonal parts of A, respectively:

$$
V_+ = \begin{pmatrix} 0 & a_{12} & a_{13} & a_{14} & \cdots \\ 0 & 0 & a_{23} & a_{24} & \cdots \\ 0 & 0 & 0 & a_{34} & \cdots \\ \cdot & \cdot & \cdot & \cdot & \cdots \end{pmatrix}, \quad V_- = \begin{pmatrix} 0 & 0 & 0 & 0 & \cdots \\ a_{21} & 0 & 0 & 0 & \cdots \\ a_{31} & a_{32} & 0 & 0 & \cdots \\ a_{41} & a_{42} & a_{43} & 0 & \cdots \\ \cdot & \cdot & \cdot & \cdots \end{pmatrix}
$$

(1.1)

and

$$D = diag\,[a_{11}, a_{22}, a_{33}, ...].$$

Throughout this chapter it is assumed that V_- and V_+ are the Hille-Tamarkin matrices. That is, for some finite $p > 1$,

$$\sum_{j=1}^{\infty} [\sum_{k=1,\ k\neq j}^{\infty} |a_{jk}|^q]^{p/q}]^{1/p} < \infty \tag{1.2}$$

with

$$\frac{1}{p} + \frac{1}{q} = 1.$$

As usually l^p $(1 < p < \infty)$ is the Banach space of number sequences equipped with the norm

$$|h|_{l^p} = [\sum_{k=1}^{\infty} |h_k|^p]^{1/p} \ (h = (h_k) \in l^p).$$

So under (1.2), A represents a linear operator in l^p which is also denoted by A. Clearly,

$$Dom\ (A) = Dom\ (D) = \{h = (h_k) \in l^p : \sum_{k=1}^{\infty} |a_{kk}h_k|^p < \infty\}.$$

Assume that

$$d_0 \equiv \inf_k |a_{kk}| > 0 \tag{1.3}$$

and introduce the notations

$$M_p^+(A) \equiv (\sum_{j=1}^{\infty} [\sum_{k=j+1}^{\infty} |a_{jj}^{-1} a_{jk}|^q]^{p/q})^{1/p},$$

$$M_p^-(A) = (\sum_{j=2}^{\infty} [\sum_{k=1}^{j-1} |a_{jj}^{-1} a_{jk}|^q]^{p/q})^{1/p},$$

and

$$J_p^{\pm}(A) = \sum_{k=0}^{\infty} \frac{(M_p^{\pm}(A))^k}{\sqrt[p]{k!}}.$$

Now we are in a position to formulate the main result of the chapter.

Theorem 18.1.1 *Let the conditions (1.2), (1.3) and*

$$J_p^-(A)J_p^+(A) < J_p^+(A) + J_p^-(A) \tag{1.4}$$

hold. Then A is boundedly invertible in l^p and the inverse operator satisfies the inequality

$$|A^{-1}|_{l^p} \leq \frac{J_p^-(A)J_p^+(A)}{(J_p^+(A) + J_p^-(A) - J_p^-(A)J_p^+(A))d_0} \tag{1.5}$$

The proof of this theorem is presented in the next section.

18.2   Proof of Theorem 18.1.1

Lemma 18.2.1 *Under condition (1.2), for the strictly upper and lower tri-angular matrices V_+ and V_-, the inequalities*

$$|V_\pm^m|_{lp} \leq \frac{(v_p^\pm)^m}{\sqrt[p]{m!}} \quad (m = 1, 2, ...)$$

are valid, where

$$v_p^+ = (\sum_{j=1}^{\infty} [\sum_{k=j+1}^{\infty} |a_{jk}|^q]^{p/q})^{1/p}$$

and

$$v_p^- = (\sum_{j=2}^{\infty} [\sum_{k=1}^{j-1} |a_{jk}|^q]^{p/q})^{1/p}.$$

This result follows from Lemma 3.2.1 when $n \to \infty$, since

$$\gamma_{n,m,p} \leq \frac{1}{\sqrt[p]{m!}}.$$

So operators $V_\pm$ are quasinilpotent. The latter lemma yields

Corollary 18.2.2 *Under conditions (1.2), (1.3), the inequalities*

$$|(D^{-1}V_\pm)^m|_{lp} \leq \frac{(M_p^\pm(A))^m}{\sqrt[p]{m!}} \quad (m = 1, 2, ...)$$

are valid.

Proof of Theorem 18.1.1: We have

$$A = V_+ + V_- + D = D(D^{-1}V_+ + D^{-1}V_- + I)$$

Clearly,

$$|D^{-1}|_{lp} = d_0^{-1}.$$

From Lemma 18.2.2, it follows that

$$j(D^{-1}V_\pm) \leq J_p^\pm(A)$$

Now Corollary 16.2.2 and condition (1.4) yield the invertibility of the operator

$$D^{-1}V_+ + D^{-1}V_- + I$$

and the estimate (1.5). □

18.3   Localization of the Spectrum

Let $\sigma(A)$ be the spectrum of A. For a $\lambda \in \mathbf{C}$, assume that

$$\rho(D, \lambda) \equiv \inf_m |\lambda - a_{mm}| > 0,$$

and put

$$M_p^+(A, \lambda) = (\sum_{j=1}^{\infty} [\sum_{k=j+1}^{\infty} |(a_{jj} - \lambda)^{-1} a_{jk}|^q]^{p/q})^{1/p},$$

$$M_p^-(A, \lambda) = (\sum_{j=2}^{\infty} [\sum_{k=1}^{j-1} |(a_{jj} - \lambda)^{-1} a_{jk}|^q]^{p/q})^{1/p},$$

and

$$J_p^\pm(A, \lambda) = \sum_{k=0}^{\infty} \frac{(M_p^\pm(A, \lambda))^k}{\sqrt[p]{k!}}.$$

Clearly

$$M_p^\pm(A, 0) = M_p^\pm(A), \ J_p^\pm(A, 0) = J_p^\pm(A).$$

Lemma 18.3.1 *Under condition (1.2), for any $\mu \in \sigma(A)$ we have either $\mu = a_{jj}$ for some natural j, or*

$$J_p^-(A, \mu) J_p^+(A, \mu) \geq J_p^+(A, \mu) + J_p^-(A, \mu). \tag{3.1}$$

Proof: Assume that

$$J_p^-(A, \mu) J_p^+(A, \mu) < J_p^+(A, \mu) + J_p^-(A, \mu)$$

for some $\mu \in \sigma(A)$. Then due to Theorem 18.1.1, $A - \mu I$ is invertible. This contradiction proves the required result. $\square$

Recall that $v_p^\pm$ are defined in Section 18.2 and denote,

$$F_p^\pm(z) = \sum_{k=0}^{\infty} \frac{(v_p^\pm)^k}{z^k \sqrt[p]{k!}} \ (z > 0) \tag{3.2}$$

Lemma 18.3.2 *Under condition (1.2), for any $\mu \in \sigma(A)$, either there is an integer m, such that, $\mu = a_{mm}$, or*

$$F_p^-(\rho(D, \mu)) F_p^+(\rho(D, \mu)) \geq F_p^-(\rho(D, \mu)) + F_p^+(\rho(D, \mu)). \tag{3.3}$$

Proof: Let $\mu \neq a_{kk}$ for all natural k. Then

$$M_p^\pm(A, \mu) \leq \rho^{-1}(D, \mu) v_p^\pm.$$

Hence,

$$J_p^\pm(A, \mu) \leq F_p^\pm(\rho(D, \mu)). \tag{3.4}$$

In addition, (3.1) is equivalent to the relation

$$(J_p^-(A, \mu) - 1)(J_p^+(A, \mu) - 1) \geq 1.$$

Now (3.4) implies (3.3). □

Theorem 18.3.3 *Under condition (1.2), let $V_+ \neq 0, V_- \neq 0$. Then the equation*

$$F_p^-(z)F_p^+(z) = F_p^-(z) + F_p^+(z) \tag{3.5}$$

has a unique positive root $\zeta(A)$. Moreover, $\rho(D, \mu) \leq \zeta(A)$ for any $\mu \in \sigma(A)$. In other words, $\sigma(A)$ lies in the closure of the union of the discs

$$\{\lambda \in \mathbf{C} : |\lambda - a_{kk}| \leq \zeta(A)\} \quad (k = 1, 2, ...).$$

Proof: Equation (3.5) is equivalent to the following one:

$$(F_p^-(z) - 1)(F_p^+(z) - 1) = 1. \tag{3.6}$$

The left part of this equation monotonically decreases as $z > 0$ increases; so it has a unique positive root $\zeta(A)$. In addition, (3.3) is equivalent to the relation

$$(F_p^-(\rho(D, \mu)) - 1)(F_p^+(\rho(D, \mu)) - 1) \geq 1. \tag{3.7}$$

Hence the result follows. □

Rewrite (3.5) as

$$\sum_{k=1}^{\infty} \frac{(v_p^-)^k}{z^k \sqrt[p]{k!}} \sum_{j=1}^{\infty} \frac{(v_p^+)^j}{z^j \sqrt[p]{j!}} = 1.$$

Or

$$\sum_{k=2}^{\infty} B_k z^k = 1 \text{ with } B_k = \sum_{j=1}^{k-1} \frac{(v_p^+)^{k-j}(v_p^-)^j}{\sqrt[p]{j!(k-j)!}} \quad (k = 2, 3, ...).$$

Due to the Lemma 8.3.1, with the notation

$$\delta_p(A) \equiv 2 \sup_{j=2,3,...} \sqrt[j]{B_j},$$

we get $\zeta(A) \leq \delta_p(A)$. Now Theorem 18.3.3 yields

Corollary 18.3.4 *Under condition (1.2), let $V_+ \neq 0, V_- \neq 0$. Then for any $\mu \in \sigma(A)$, the inequality $\rho(\mu, D) \leq \delta(A)$ is true.*

In other words, $\sigma(A)$ lies in the closure of the union of the sets

$$\{\lambda \in \mathbf{C} : |\lambda - a_{kk}| \leq \delta_p(A)\} \quad (k = 1, 2, ...).$$

Note that Theorem 18.3.3 is exact: if A is triangular: either $V_- = 0$, or $V_+ = 0$, then we due to that lemma $\sigma(A)$ is the closure of the set

$$\{a_{kk}, \ k = 1, 2, ...\}.$$

Moreover, Theorem 18.3.3 and Corollary 18.3.4 imply

$$r_s(A) \leq \sup_{k=1,2,...} |a_{kk}| + \zeta(A) \leq \sup_{k=1,2,...} |a_{kk}| + \delta_p(A), \tag{3.8}$$

provided D is bounded. Furthermore, let the condition

$$\sup_{j=1,2,...} \sum_{k=1}^{\infty} |a_{jk}| < \infty \tag{3.9}$$

hold. Then the well-known estimate

$$r_s(A) \leq \sup_{j=1,2,...} \sum_{k=1}^{\infty} |a_{jk}| \tag{3.10}$$

is valid, see (Krasnosel'skii et al. 1989, Theorem 16.2). Under condition (3.9), relations (3.8) improve (3.10), provided

$$\zeta(A) < \sup_{j=1,2,...} \sum_{k=1, k \neq j}^{\infty} |a_{jk}|$$

or

$$\delta_p(A) < \sup_{j=1,2,...} \sum_{k=1, k \neq j}^{\infty} |a_{jk}|.$$

In conclusion, note that Theorem 18.1.1 is exact: if A is upper or lower triangular, then A is invertible, provided D is invertible.

18.4   Notes

The present chapter is based on the paper (Gil', 2002).

About other results on the spectrum of Hille-Tamarkin matrices see, for instance, the books (Diestel et al., 1995), (König, 1986), (Pietsch, 1987), and references therein.

Note that Hille-Tamarkin matrices arise, in particular, in recent investigations of discrete Volterra equations, see (Kolmanovskii et al, 2000), (Gil' and Medina, 2002), (Medina and Gil', 2003).

References

[1] Diestel, D., Jarchow, H, Tonge, A. (1995), *Absolutely Summing Operators*, Cambridge University Press, Cambridge.

[2] Gil', M.I. (2002), Invertibility and spectrum of Hille-Tamarkin matrices, *Mathematische Nachrichten*, **244**, 1-11

[3] Gil', M.I. and Medina, R. (2002). Boundedness of solutions of matrix nonlinear Volterra difference equations. *Discrete Dynamics in Nature and Society*, **7**, No 1, 19-22

[4] Kolmanovskii, V.B., A.D. Myshkis and J.P. Richard (2000). Estimate of solutions for some Volterra difference equations, *Nonlinear Analysis, TMA*, **40**, 345-363.

[5] König, H. (1986). *Eigenvalue Distribution of Compact Operators*, Birkhäuser Verlag, Basel- Boston-Stuttgart.

[6] Krasnosel'skii, M. A., J. Lifshits, and A. Sobolev (1989). *Positive Linear Systems. The Method of Positive Operators*, Heldermann Verlag, Berlin.

[7] Medina, R. and Gil', M.I. (2003). Multidimensional Volterra difference equations. In the book: *New Progress in Difference Equations*, Eds. S. Elaydi, G. Ladas and B. Aulbach, Taylor and Francis, London and New York, p. 499-504

[8] Pietsch, A. (1987). *Eigenvalues and s-Numbers*, Cambridge University Press, Cambridge.

19. Zeros of Entire Functions

The present chapter is devoted to applications of our abstract results to the theory of finite order entire functions. We consider the following problem: if the Taylor coefficients of two entire functions are close, how close are their zeros? In addition, we establish bounds for sums of the absolute values of the zeros in the terms of the coefficients of its Taylor series. These bounds supplement the Hadamard theorem.

19.1   Perturbations of Zeros

Consider the entire function

$$f(\lambda) = \sum_{k=0}^{\infty} c_k \lambda^k \ (\lambda \in \mathbf{C}; \ c_0 = 1)$$

with complex, in general, coefficients $c_k, k = 1, 2, \dots$. Put

$$M_f(r) := \max_{|z|=r} |f(z)| \ (r > 0).$$

Recall that

$$\rho\,(f) := \overline{\lim}_{r \to \infty} \frac{\ln \ln M_f(r)}{\ln r}$$

is the order of f. Moreover, the relation

$$\rho\,(f) = \overline{\lim}_{n \to \infty} \frac{n \ln n}{\ln \,(1/|c_n|)}$$

is true, cf. (Levin, 1996, p. 6).

Everywhere in the present chapter it is assumed that the set

$$\{z_k(f)\}_{k=1}^{\infty}$$

of all the zeros of f taken with their multiplicities is infinite.

Note that if f has a finite number m of the zeros, we can put $z_k^{-1}(f) = 0$ for $k = m, m+1, \ldots$ and aplly our arguments below. Here and below $z_k^{-1}(f)$ means $\frac{1}{z_k(f)}$.

Rewrite function f in the form

$$f(\lambda) = \sum_{k=0}^{\infty} \frac{a_k \lambda^k}{(k!)^{\gamma}} \quad (a_0 = 1) \tag{1.1a}$$

with a positive γ, and consider the function

$$h(\lambda) = \sum_{k=0}^{\infty} \frac{b_k \lambda^k}{(k!)^{\gamma}} \quad (b_0 = 1). \tag{1.1b}$$

Assume that

$$\sum_{k=0}^{\infty} |a_k|^2 < \infty, \ \sum_{k=0}^{\infty} |b_k|^2 < \infty. \tag{1.2}$$

Relations (1.1) and (1.2), imply that functions f and h have orders no more than $1/\gamma$.

Definition 19.1.1 *The quantity*

$$zv_f(h) = \max_j \min_k |z_k^{-1}(f) - z_j^{-1}(h)|$$

will be called the variation of zeros of function h with respect to function f.

For a natural $p > 1/2\gamma$, put

$$w_p(f) := 2 \left[\sum_{k=1}^{\infty} |a_k|^2\right]^{1/2} + 2 \left[\zeta(2\gamma p) - 1\right]^{1/2p}, \tag{1.3}$$

where ζ is the Riemann Zeta function, and

$$\psi(f, y) := \sum_{k=0}^{p-1} \frac{w^k(f)}{y^{k+1}} exp \left[\frac{1}{2} + \frac{w_p^{2p}(f)}{2y^{2p}}\right] \quad (y > 0). \tag{1.4}$$

Finally, denote

$$q := \left[\sum_{k=1}^{\infty} |a_k - b_k|^2\right]^{1/2}.$$

Theorem 19.1.2 *Let conditions (1.1) and (1.2) be fulfilled. In addition, let $r(q, f)$ be the unique positive (simple) root of the equation*

$$q\psi(f, y) = 1.$$

Then $zv_f(h) \leq r(q, f)$. That is, for any zero $z(h)$ of h there is a zero $z(f)$ of f, such that

$$|z(h) - z(f)| \leq r(q, f)|z(h)z(f)|. \tag{1.5}$$

The proof of Theorem 19.1.2 is presented in the next section. Substitute in (1.5) the equality $y = xw_p(f)$ and apply Lemma 8.3.2. Then we have

$$r(q, f) \leq \delta(q, f), \tag{1.6}$$

where

$$\delta(q, f) := \begin{cases} epq & \text{if } w_p(f) \leq epq, \\ w_p(f) \left[\ln\left(w_p(f)/qp\right)\right]^{-1/2p} & \text{if } w_p(f) > epq \end{cases}.$$

Theorem 19.1.2 and inequality (1.6) yield

Corollary 19.1.3 *Let conditions (1.1) and (1.2) be fulfilled. Then $zv_f(h) \leq \delta(q, f)$. That is, for any zero $z(h)$ of h, there is a zero $z(f)$ of f, such that*

$$|z(h) - z(f)| \leq \delta(q, f)|z(h)z(f)|. \tag{1.7}$$

Relations (1.5) and (1.7) imply the inequalities

$$|z(f)| - |z(h)| \leq r(q, f)|z(h)||z(f)| \leq \delta(q, f)|z(h)||z(f)|.$$

Hence,

$$|z(h)| \geq (r(q, f)|z(f)| + 1)^{-1}|z(f)| \geq (\delta(q, f)|z(f)| + 1)^{-1}|z(f)|.$$

This inequality yields the following result

Corollary 19.1.4 *Under conditions (1.1) and (1.2), for a positive number R_0, let f have no zeros in the disc $\{z \in \mathbf{C} : |z| \leq R_0\}$. Then h has no zeros in the disc $\{z \in \mathbf{C} : |z| \leq R_1\}$ with*

$$R_1 = \frac{R_0}{\delta(q, f)R_0 + 1} \quad or \quad R_1 = \frac{R_0}{r(q, f)R_0 + 1}.$$

Let us assume that under (1.1), there is a constant $d_0 \in (0, 1)$, such that

$$\overline{\lim}_{k \to \infty} \sqrt[k]{|a_k|} < 1/d_0 \tag{1.8}$$

and

$$\overline{\lim}_{k \to \infty} \sqrt[k]{|b_k|} < 1/d_0$$

and consider the functions

$$\tilde{f}(\lambda) = \sum_{k=0}^{\infty} \frac{a_k(d_0\lambda)^k}{(k!)^\gamma} \tag{1.9}$$

and

$$\tilde{h}(\lambda) = \sum_{k=0}^{\infty} \frac{b_k(d_0\lambda)^k}{(k!)^\gamma}.$$

That is, $\tilde{f}(\lambda) \equiv f(d_0\lambda)$ and $\tilde{h}(\lambda) \equiv h(d_0\lambda)$. So functions $\tilde{f}(\lambda)$ and $\tilde{h}(\lambda)$ satisfy conditions (1.2). Moreover,

$$w_p(\tilde{f}) = 2[\sum_{k=1}^{\infty} d_0^{2k}|a_k|^2]^{1/2} + 2[\zeta(2\gamma p) - 1]^{1/2p}.$$

Thus, we can apply Theorem 19.1.2 and its corollaries to functions $\tilde{f}(\lambda), \tilde{h}(\lambda)$ and take into account that

$$d_0 z_k(\tilde{f}) = z_k(f), \ d_0 z_k(\tilde{h}) = z_k(h). \tag{1.10}$$

19.2   Proof of Theorem 19.1.2

For a finite integer n, consider the polynomials

$$F(\lambda) = \sum_{k=0}^{n} \frac{a_k\lambda^{n-k}}{(k!)^\gamma} \ \text{ and } \ Q(\lambda) = \sum_{k=0}^{n} \frac{b_k\lambda^{n-k}}{(k!)^\gamma} \ \ (a_0 = b_0 = 1). \tag{2.1}$$

Put

$$w_p(F) := 2\left[\sum_{k=1}^{n} |a_k|^2\right]^{1/2} + 2\left[\sum_{k=2}^{n} 1/k^{2\gamma p}\right]^{1/2p} \text{ and } q(F,Q) := \left[\sum_{k=1}^{n} |a_k - b_k|^2\right]^{1/2}.$$

In addition, $\{z_k(F)\}_{k=1}^{n}$ and $\{z_k(Q)\}_{k=1}^{n}$ are the sets of all the zeros of F and Q, respectively taken with their multiplicities. Define $\psi(F, y)$ according to (1.4).

Lemma 19.2.1 *For any zero $z(Q)$ of Q, there is a zero $z(F)$ of F, such that*

$$|z(F) - z(Q)| \le r(Q, F),$$

where $r(Q, F)$ be the unique positive (simple) root of the equation

$$q(F, Q)\psi(F, y) = 1. \tag{2.2}$$

Proof: In a Euclidean space $\mathbf{C}^n$ with the Euclidean norm $\|.\|$, introduce operators A_n and B_n by virtue of the $n \times n$-matrices

$$A_n = \begin{pmatrix} -a_1 & -a_2 & \dots & -a_{n-1} & -a_n \\ 1/2^\gamma & 0 & \dots & 0 & 0 \\ 0 & 1/3^\gamma & \dots & 0 & 0 \\ . & . & \dots & . & . \\ 0 & 0 & \dots & 1/n^\gamma & 0 \end{pmatrix}$$

and

$$B_n = \begin{pmatrix} -b_1 & -b_2 & \dots & -b_{n-1} & -b_n \\ 1/2^\gamma & 0 & \dots & 0 & 0 \\ 0 & 1/3^\gamma & \dots & 0 & 0 \\ . & . & \dots & . & . \\ 0 & 0 & \dots & 1/n^\gamma & 0 \end{pmatrix}.$$

It is simple to see that

$$F(\lambda) = det\,(\lambda I - A_n)$$

and $Q(\lambda) = det\,(\lambda I - B_n)$, where I is the unit matrix. So

$$\lambda_k(A_n) = z_k(F),\ \lambda_k(B_n) = z_k(Q) \quad (k = 1, 2, ..., n), \tag{2.3}$$

where $\lambda_k(.), k = 1, ..., n$ are the eigenvalues with their multiplicities. Clearly,

$$\|A_n - B_n\| = q(F, Q).$$

Due to Theorem 8.5.4, for any $\lambda_j(B_n)$ there is an $\lambda_i(A_n)$, such that

$$|\lambda_j(B_n) - \lambda_i(A_n)| \le y_p(A_n, B_n), \tag{2.4}$$

where $y_p(A_n, B_n)$ is the unique positive (simple) root of the equation

$$q(F, Q) \sum_{k=0}^{p-1} \frac{(2N_{2p}(A_n))^k}{y^{k+1}} exp\,[(1 + \frac{(2N_{2p}(A_n))^{2p}}{y^{2p}})/2] = 1,$$

where $N_{2p}(A) := [Trace(AA^*)^p]^{1/2p}$ is the Neumann-Schatten norm and the asterisk means the adjointness. But $A_n = M + C$, where

$$M = \begin{pmatrix} -a_1 & -a_2 & \dots & -a_{n-1} & -a_n \\ 0 & 0 & \dots & 0 & 0 \\ . & . & \dots & . & . \\ 0 & 0 & \dots & 0 & 0 \end{pmatrix}$$

and

$$C = \begin{pmatrix} 0 & 0 & \dots & 0 & 0 \\ 1/2^\gamma & 0 & \dots & 0 & 0 \\ 0 & 1/3^\gamma & \dots & 0 & 0 \\ . & . & \dots & . & . \\ 0 & 0 & \dots & 1/n^\gamma & 0 \end{pmatrix}.$$

Therefore, with

$$c = \sum_{k=1}^{n} |a_k|^2,$$

we have

$$MM^* = \begin{pmatrix} c & 0 & ... & 0 & 0 \\ 0 & 0 & ... & 0 & 0 \\ . & . & ... & . & . \\ 0 & 0 & ... & 0 & 0 \end{pmatrix}$$

and

$$CC^* = \begin{pmatrix} 0 & 0 & ... & 0 & 0 \\ 0 & 1/2^{2\gamma} & ... & 0 & 0 \\ 0 & 0 & ... & 0 & 0 \\ . & . & ... & . & . \\ 0 & 0 & ... & 0 & 1/n^{2\gamma} \end{pmatrix}.$$

Hence,

$$N_{2p}(A_n) \leq N_{2p}(M) + N_{2p}(C) = \sqrt{c} + [\sum_{k=2}^{n} 1/k^{2\gamma p}]^{1/2p}.$$

Consequently $y_p(A_n, B_n) \leq r(Q, F)$. Therefore (2.3) and (2.4) imply (2.2), as claimed. $\square$

Proof of Theorem 19.1.2: Consider the polynomials

$$f_n(\lambda) = \sum_{k=0}^{n} \frac{a_k \lambda^k}{(k!)^\gamma} \quad \text{and} \quad h_n(\lambda) = \sum_{k=0}^{n} \frac{b_k \lambda^k}{(k!)^\gamma}. \tag{2.5}$$

Clearly, $\lambda^n f_n(1/\lambda) = F(\lambda)$ and $h_n(1/\lambda)\lambda^n = Q(\lambda)$. So

$$z_k(F) = 1/z_k(f_n); \quad z_k(Q) = 1/z_k(h_n). \tag{2.6}$$

Take into account that the roots continuously depend on coefficients, we have the required result, letting in the previous lemma $n \to \infty$. $\square$

19.3   Bounds for Sums of Zeros

Again consider an entire function f of the form (1.1a) and assume that the condition

$$\theta_f := [\sum_{k=1}^{\infty} |a_k|^2]^{1/2} < \infty \tag{3.1}$$

holds. Let the zeros of f be numerated in the increasing way:

$$|z_k(f)| \leq |z_{k+1}(f)| \quad (k = 1, 2, ...). \tag{3.2}$$

Theorem 19.3.1 *Let f be an entire function of the form (1.1a). Then under conditions (3.1) and (3.2), the inequalities*

$$\sum_{k=1}^{j} |z_k(f)|^{-1} \leq \theta_f + \sum_{k=1}^{j} (k+1)^{-\gamma} \ (j=1,2,...)$$

are valid.

The proof of this theorem is presented in this section below. Note that under condition (1.8) we can omit condition (3.1) due to (1.9) and (1.10).

To prove Theorem 19.3.1, again consider the polynomial $F(\lambda)$ defined in (2.1) with the zeros ordered in the following way:

$$|z_k(F)| \geq |z_{k+1}(F)| \ (k=1,...,n-1).$$

Set

$$\theta(F) := [\sum_{k=1}^{n} |a_k|^2]^{1/2}.$$

Lemma 19.3.2 *The zeros of F satisfy the inequalities*

$$\sum_{k=1}^{j} |z_k(F)| \leq \theta(F) + \sum_{k=1}^{j} (k+1)^{-\gamma} \ \ (j=1,...,n-1)$$

and

$$\sum_{k=1}^{n} |z_k(F)| \leq \theta(F) + \sum_{k=1}^{n-1} (k+1)^{-\gamma}.$$

Proof: Take into account that according to (2.3),

$$\sum_{k=1}^{j} |\lambda_k(A_n)| \leq \sum_{k=1}^{j} s_k(A_n) \ (j=1,...,n), \tag{3.3}$$

where $s_k(A_n), k = 1, 2, ...$ are the singular numbers of A_n ordered in the decreasing way (Marcus and Minc, 1964, Section II.4.2). But $A_n = M + C$, where M and C are introduced in Section 19.2. We can write

$$s_1(M) = \theta(F), \ s_k(M) = 0 \ (k=2,...,n).$$

In addition,

$$s_k(C) = 1/(k+1)^{\gamma} \ (k=1,...,n-1), \ s_n(C) = 0.$$

Take into account that

$$\sum_{k=1}^{j} s_k(A_n) = \sum_{k=1}^{j} s_k(M+C) \leq \sum_{k=1}^{j} s_k(M) + \sum_{k=1}^{j} s_k(C),$$

cf. (Gohberg and Krein, 1969, Lemma II.4.2). So

$$\sum_{k=1}^{j} s_k(A_n) \leq \theta(F) + \sum_{k=1}^{j}(k+1)^{-\gamma} \quad (j = 1, ..., n-1)$$

and

$$\sum_{k=1}^{n} s_k(A_n) \leq \theta(F) + \sum_{k=1}^{n-1}(k+1)^{-\gamma}.$$

Now (2.3) and (3.3) yield the required result. $\square$

Proof of Theorem 19.3.1:   Again consider the polynomial $f_n(z)$ defined as in (2.5). Now Lemma 19.3.2 and (2.6) yield the inequalities

$$\sum_{k=1}^{j} |z_k(f_n)|^{-1} \leq \theta_f + \sum_{k=1}^{j}(k+1)^{-\gamma} \quad (j = 1, ..., n-1). \tag{3.4}$$

But the zeros of entire functions continuously depend on its coefficients. So for any $j = 1, 2, ...,$

$$\sum_{k=1}^{j} |z_k(f_n)|^{-1} \to \sum_{k=1}^{j} |z_k(f)|^{-1}$$

as $n \to \infty$. Now (3.4) implies the required result. $\square$

19.4   Applications of Theorem 19.3.1

Put

$$\tau_1 = \theta_f + 2^{-\gamma} \text{ and } \tau_k = (k+1)^{-\gamma} \ (k = 2, 3, ...).$$

Corollary 19.4.1 *Let $\phi(t)$ $(0 \leq t < \infty)$ be a convex scalar-valued function, such that $\phi(0) = 0$. Then under conditions (1.1a), (3.1) and (3.2), the inequalities*

$$\sum_{k=1}^{j} \phi(|z_k(f)|^{-1}) \leq \sum_{k=1}^{j} \phi(\tau_k) \ (j = 1, 2, ...)$$

are valid. In particular, for any $r \geq 2$,

$$\sum_{k=1}^{j} |z_k(f)|^{-r} \leq \sum_{k=1}^{j} \tau_k^r = (\theta_f + 2^{-\gamma})^r + \sum_{k=2}^{j}(k+1)^{-r\gamma} \ (j = 2, 3, ...). \tag{4.1}$$

Indeed, this result is due to the well-known Lemma II.3.4 (Gohberg and Krein, 1969) and Theorem 19.3.1.

Furthermore, assume that

$$r\gamma > 1, \ r \geq 2. \tag{4.2}$$

Then the series

$$\sum_{k=1}^{\infty} \tau_k^r = (\theta_f + 2^{-\gamma})^r + \sum_{k=2}^{\infty} (k+1)^{-r\gamma} = (\theta_f + 2^{-\gamma})^r + \zeta(\gamma r) - 1 - 2^{-r\gamma}$$

converges. Here $\zeta(.)$ is the Riemann Zeta function, again. Now relation (4.1) yields

Corollary 19.4.2 *Under the conditions (1.1a), (3.1) and (4.2), the inequality*

$$\sum_{k=1}^{\infty} |z_k(f)|^{-r} \leq (\theta_f + 2^{-\gamma})^r + \zeta(\gamma r) - 1 - 2^{-\gamma r} \tag{4.3}$$

is valid. In particular, if $\gamma > 1$, then due to (3.4)

$$\sum_{k=1}^{\infty} |z_k(f)|^{-1} \leq \theta_f + \zeta(\gamma) - 1. \tag{4.4}$$

Consider now a positive scalar-valued function $\Phi(t_1, t_2, ..., t_j)$ with an integer j, defined on the domain

$$0 \leq t_j \leq t_{j-1} \leq t_2 \leq t_1 < \infty$$

and satisfying

$$\frac{\partial \Phi}{\partial t_1} > \frac{\partial \Phi}{\partial t_2} > ... > \frac{\partial \Phi}{\partial t_j} > 0 \ \text{for} \ t_1 > t_2 > ... > t_j. \tag{4.5}$$

Corollary 19.4.3 *Under conditions (1.1a), (3.1), (3.2) and (4.5),*

$$\Phi(|z_1(f)|^{-1}, |z_2(f)|^{-1}, ..., |z_j(f)|^{-1}) \leq \Phi(\tau_1, \tau_2, ..., \tau_j).$$

Indeed, this result is due to Theorem 19.3.1 and the well-known Lemma II.3.5 (Gohberg and Krein, 1969).

In particular, let $\{d_k\}_{k=1}^{\infty}$ be a decreasing sequence of non-negative numbers. Take

$$\Phi(t_1, t_2, ..., t_j) = \sum_{k=1}^{j} d_k t_k.$$

Then Corollary 19.4.3 yields

$$\sum_{k=1}^{j} d_k |z_k(f)|^{-1} \leq \sum_{k=1}^{j} \tau_k d_k = d_1 \theta_f + \sum_{k=1}^{j} d_k (k+1)^{-\gamma}$$

$$(j = 2, 3, ...).$$

19.5   Notes

The variation of the zeros of general analytic functions under perturbations was investigated, in particular, by P. Rosenbloom (1969). He established the perturbation result that provides the existence of a zero of a perturbed function in a given domain. In the present chapter a new approach to the problem is proposed.

The material in the present chapter is taken from the papers (Gil', 2000a, 2000b, 2000c and 2001). Corollary 19.4.2 supplements the classical Hadamard theorem (Levin, 1996, p. 18), since it not only asserts the convergence of the series of the zeros, but also gives us the estimate for the sums of the zeros.

References

[1] Gil', M.I. (2000a). Inequalities for imaginary parts of zeros of entire functions. *Results in Mathematics*, **37**, 331-334

[2] Gil', M.I. (2000b). Perturbations of zeros of a class of entire functions, *Complex Variables*, **42**, 97-106

[3] Gil', M.I. (2000c). Approximations of zeros of entire functions by zeros of polynomials. *J. of Approximation Theory*, **106**, 66-76

[4] Gil', M.I. (2001). Inequalities for zeros of entire functions, *Journal of Inequalities*, **6** 463-471.

[5] Gohberg, I. C. and Krein, M. G. (1969). *Introduction to the Theory of Linear Nonselfadjoint Operators*, Trans. of Math. Monographs, v. 18, Amer. Math. Soc., R.I.

[6] Levin, B. Ya. (1996). *Lectures on Entire Functions*, Trans. of Math. Monographs, v. 150. Amer. Math. Soc., R. I.

[7] Marcus, M. and Minc, H. (1964). *A Survey of Matrix Theory and Matrix Inequalities*, Allyn and Bacon, Boston.

[8] Rosenbloom, P.C. (1969). Perturbation of zeros of analytic functions. I. *Journal of Approximation Theory*, **2**, 111-126.

List of Main Symbols

Index

Vol. 1797: B. Schmidt, Characters and Cyclotomic Fields in Finite Geometry. VIII, 100 pages. 2002.

Vol. 1798: W.M. Oliva, Geometric Mechanics. XI, 270 pages. 2002.

Vol. 1799: H. Pajot, Analytic Capacity, Rectifiability, Menger Curvature and the Cauchy Integral. XII,119 pages. 2002.

Vol. 1800: O. Gabber, L. Ramero, Almost Ring Theory. VI, 307 pages. 2003.

Vol. 1801: J. Azéma, M. Émery, M. Ledoux, M. Yor, Séminaire de Probabilités XXXVI. VIII, 499 pages. 2003.

Vol. 1802: V. Capasso, E. Merzbach, B.G. Ivanoff, M. Dozzi, R. Dalang, T. Mountford, Topics in Spatial Stochastic Processes. Martina Franca, Italy 2001. Editor: E. Merzbach. VIII, 253 pages. 2003.

Vol. 1803: G. Dolzmann, Variational Methods for Crystalline Microstructure - Analysis and Computation. VIII, 212 pages. 2003.

Vol. 1804: I. Cherednik, Ya. Markov, R. Howe, G. Lusztig, Iwahori-Hecke Algebras and their Representation Theory. Martina Franca, Italy 1999. Editors: V. Baldoni, D. Barbasch. X, 103 pages. 2003.

Vol. 1805: F. Cao, Geometric Curve Evolution and Image Processing. X, 187 pages. 2003.

Vol. 1806: H. Broer, I. Hoveijn. G. Lunther, G. Vegter, Bifurcations in Hamiltonian Systems. Computing Singularities by Gröbner Bases. XIV, 169 pages. 2003.

Vol. 1807: V. D. Milman, G. Schechtman, Geometric Aspects of Functional Analysis. Israel Seminar 2000-2002. VIII, 429 pages. 2003.

Vol. 1808: W. Schindler, Measures with Symmetry Properties.IX, 167 pages. 2003.

Vol. 1809: O. Steinbach, Stability Estimates for Hybrid Coupled Domain Decomposition Methods. VI, 120 pages. 2003.

Vol. 1810: J. Wengenroth, Derived Functors in Functional Analysis. VIII, 134 pages. 2003.

Vol. 1811: J. Stevens, Deformations of Singularities. VII, 157 pages. 2003.

Vol. 1812: L. Ambrosio, K. Deckelnick, G. Dziuk, M. Mimura, V. A. Solonnikov, H. M. Soner, Mathematical Aspects of Evolving Interfaces. Madeira, Funchal, Portugal 2000. Editors: P. Colli, J. F. Rodrigues. X, 237 pages. 2003.

Vol. 1813: L. Ambrosio, L. A. Caffarelli, Y. Brenier, G. Buttazzo, C. Villani, Optimal Transportation and its Applications. Martina Franca, Italy 2001. Editors: L. A. Caffarelli, S. Salsa. X, 164 pages. 2003.

Vol. 1814: P. Bank, F. Baudoin, H. Föllmer, L.C.G. Rogers, M. Soner, N. Touzi, Paris-Princeton Lectures on Mathematical Finance. X,172 pages. 2003.

Vol. 1815: A. M. Vershik (Ed.), Asymptotic Combinatorics with Applications to Mathematical Physics. St. Petersburg, Russia 2001. IX, 246 pages. 2003.

Vol. 1816: S. Albeverio, W. Schachermayer, M. Talagrand, Lectures on Probability Theory and Statistics. Ecole d'Eté de Probabilités de Saint-Flour XXX-2000. Editor: P. Bernard. VIII, 296 pages. 2003.

Vol. 1817: E. Koelink (Ed.), Orthogonal Polynomials and Special Functions. Leuven 2002. X, 249 pages. 2003.

Vol. 1818: M. Bildhauer, Convex Variational Problems with Linear, nearly Linear and/or Anisotropic Growth Conditions. X, 217 pages. 2003.

Vol. 1819: D. Masser, Yu. V. Nesterenko, H. P. Schlickewei, W. M. Schmidt, M. Waldschmidt, Diophantine Approximation. Cetraro, Italy 2000. Editors: F. Amoroso, U. Zannier. XI,353 pages. 2003.

Vol. 1820: F. Hiai, H. Kosaki, Means of Hilbert Space Operators. VIII, 148 pages. 2003.

Vol. 1821: S. Teufel, Adiabatic Perturbation Theory in Quantum Dynamics. VI, 242 pages. 2003.

Vol. 1822: S.-N. Chow, R. Conti, R. Johnson, J. Mallet-Paret, R. Nussbaum, Dynamical Systems. Cetraro, Italy 2000. Editors: J. W. Macki, P. Zecca. XII, 345 pages. 2003.

Vol. 1823: A. M. Anile, W. Allegretto, C. Ringhofer, Mathematical Problems in Semiconductor Physics. Cetraro, Italy 1998. Editor: A. M. Anile. X, 135 pages. 2003.

Vol. 1824: J. A. Navarro González, J. B. Sancho de Salas, C^∞ - Differentiable Spaces. XIII, 188 pages. 2003.

Vol. 1825: J. H. Bramble, A. Cohen, W. Dahmen, Multiscale Problems and Methods in Numerical Simulations, Martina Franca, Italy 2001. Editor: C. Canuto. XIII, 163 pages. 2003.

Vol. 1826: K. Dohmen, Improved Bonferroni Inequalities via Abstract Tubes. Inequalities and Identities of Inclusion-Exclusion Type. VIII, 113 pages, 2003.

Vol. 1827: K. M. Pilgrim, Combinations of Complex Dynamical Systems. X, 118 pages, 2003.

Vol. 1828: D. J. Green, Gröbner Bases and the Computation of Group Cohomology. XII, 138 pages, 2003.

Vol. 1829: E. Altman, B. Gaujal, A. Hordijk, Discrete-Event Control of Stochastic Networks: Multimodularity and Regularity. XIV, 313 pages, 2003.

Vol. 1830: M. I. Gil', Operator Functions and Localization of Spectra. XIV, 256 pages, 2003.

Vol. 1831: A. Connes, J. Cuntz, E. Guentner, N. Higson, J. E. Kaminker, Noncommutative Geometry, Martina Franca, Italy 2002. Editors: S. Doplicher, L. Longo. XV, 344 pages. 2003.

Recent Reprints and New Editions

Vol. 1200: V. D. Milman, G. Schechtman, Asymptotic Theory of Finite Dimensional Normed Spaces. 1986. – Corrected Second Printing. X, 156 pages. 2001.

Vol. 1471: M. Courtieu, A.A. Panchishkin, Non-Archimedean L-Functions and Arithmetical Siegel Modular Forms. – Second Edition. VII. 196 pages. 2001.

Vol. 1618: G. Pisier, Similarity Problems and Completely Bounded Maps. 1995 – Second, Expanded Edition VII, 198 pages. 2001.

Vol. 1629: J. D. Moore, Lectures on Seiberg-Witten Invariants. 1997 – Second Edition. VIII, 121 pages. 2001.

Vol. 1638: P. Vanhaecke, Integrable Systems in the realm of Algebraic Geometry. 1996 – Second Edition. X, 256 pages. 2001.

Vol. 1702: J. Ma, J. Yong, Forward-Backward Stochastic Differential Equations and Their Applications. 1999. – Corrected Second Printing. XIII, 270 pages. 2000.